高等学校中外合作办学适用教材

Differential and Difference Equations

差分与微分方程

李艳秋　郑冬梅　江舜君　编著

化学工业出版社
·北京·

This book introduces the basic solutions and theories of difference and ordinary differential equations in detail, which meets the requirements of the relevant professional syllabus and is composed of 7 chapters, including the common practical models of difference and differential equations, the solutions of difference equations, the solutions of first order and higher order differential equations, the basic theory of differential equations and the solutions of linear differential equations, qualitative theory. This book starts with not the basic concepts and theories, but the practical model and pays more attention to the application of theoretical knowledge.

This book can be used as a reference for students, teachers and researchers of mathematics, physics, engineering and related majors in colleges and universities.

本书详细介绍了差分方程和常微分方程的基本解法和基本理论，其内容符合相关专业教学大纲的要求，共由七章组成，包括常见的差分和微分方程实际模型，差分方程的求解，一阶及高阶微分方程求解，微分方程组的基本理论及线性微分方程组的解法，定性理论初步。本书并没有以基本概念和理论作为开端，而是从实际模型出发，更加注重理论知识的应用。

本书可供高等学校数学、物理、工程及相关专业的学生、教师及研究人员参考使用。

图书在版编目（CIP）数据

差分与微分方程＝Differential and Difference Equations：英文/李艳秋，郑冬梅，江舜君编著. —北京：化学工业出版社，2020.1

高等学校中外合作办学适用教材

ISBN 978-7-122-35820-2

Ⅰ.①差… Ⅱ.①李… ②郑… ③江… Ⅲ.①差分方程-高等学校-教材-英文②微分方程-高等学校-教材-英文 Ⅳ.①O241.3②O175

中国版本图书馆 CIP 数据核字（2019）第 275534 号

责任编辑：郝英华　　装帧设计：刘丽华

责任校对：宋　玮

出版发行：化学工业出版社（北京市东城区青年湖南街 13 号　邮政编码 100011）

印　　装：三河市延风印装有限公司

710mm×1000mm　1/16　印张 12¾　字数 261 千字　2020 年 2 月北京第 1 版第 1 次印刷

购书咨询：010-64518888　　售后服务：010-64518899

网　　址：http://www.cip.com.cn

凡购买本书，如有缺损质量问题，本社销售中心负责调换。

定　价：49.00 元

Preface

Ordinary differential equation is not only a compulsory basic course for mathematics, applied mathematics, financial mathematics and so on, but also an important theoretical basis for other disciplines. The purpose of this book is to provide students with the necessary theoretical knowledge and related skills.

This book introduces some commonly used practical difference and differential models to convey to readers the importance and interest of the knowledge, together with the idea of solving practical problems. The next part is used to expand the relevant theoretical system, including the basic concepts and solving methods of difference and differential equations, and the qualitative and stability theory of ordinary differential equations. In order to give students a deeper understanding of the basic theories and knowledge they have learned, and to improve their ability to solve practical problems, we provide mathematical experiments using MATLAB in the appendix.

Our book strives to be easy to understand and detailed, and the selected examples and models are as practical as possible, paying attention not only to the students' understanding of the basic concepts and methods, but also to the application of knowledge; in addition, the discussion method of this book tries to be in line with the readers' thinking habits, the mathematical expression is succinct and clear, and the professional nouns are refined and explained in the form of footnotes on each page. Easy for readers to read and find.

The Chapters 1-3 of the book are written by Dr. Dongmei Zheng, the Chapters 4-5 are written by Dr. Shunjun Jiang, and the Chapters 6-7 and Appendix are written by Dr. Yanqiu Li. Dr Yanqiu Li is responsible final compilation and editing.

Because of our limited ability, mistakes and shortcomings in the book are inevitable. I sincerely hope that readers will not be stingy with their advice.

Yanqiu Li, Dongmei Zheng, Shunjun Jiang

2019. 10

前 言

常微分方程是数学、应用数学、金融数学等专业的必修基础课程，同时也是其他学科所必需的重要理论基础。编写本书的目的是希望为学生提供必备的理论知识和相关技能。

本书为读者介绍了一些常用的微分和差分实际模型，向他们传递这部分知识的重要性和趣味性，同时也为他们提供了解决实际问题的思路；接下来依次展开相关理论体系，包括差分和微分方程、微分方程组的基本概念、求解方法，常微分方程的定性及稳定性理论初步；为了让学生更深刻地体会所学的基本理论和知识，并提高学生应用差分和微分方程解决实际问题的能力，我们在附录部分提供了利用MATLAB求解的数学实验。

本书在写法上努力做到通俗易懂，详略得当，所选实例及模型尽量做到理论联系实际，既关注学生对于基本概念和基本方法的理解，也注重知识的应用；另外本书的论述方法尽量做到符合读者的思维习惯，数学表达清晰明确，并在每一页以脚注的形式对专业名词进行了提炼和解释，便于读者阅读和查找。

本书的第一、二、三章由郑冬梅博士执笔，第四、五章由江舜君博士完成，第六、七章及附录部分由李艳秋博士负责撰写，全书由李艳秋博士统稿。

本书的编写获得“江苏省第二批中外合作办学高水平示范性建设工程项目培育点：南京工业大学与英国谢菲尔德大学合作举办数学与应用数学（金融数学）专业本科教育项目”（苏教办外［2017］14号）经费支持。

由于作者水平有限，书中不足之处在所难免，诚恳希望读者不吝赐教。

李艳秋，郑冬梅，江舜君

2019年10月

Contents

Chapter 1

Basic difference equations models[1]

Discrete phenomena exist widely in nature. The data in economics, ecology, social sciences, public health, and daily life are aggregated and counted by days, weeks, months, quarters, years, etc.. Difference equations, which are based on recurrence relations, usually describe the evolution of certain phenomena over the course of time. Difference equations are widely used in daily life and various fields. In this chapter, we will introduce some basic models leading to difference equations.

1.1 Difference equations of financial mathematics

1.1.1 Compound interest[2] and loan repayments[3]

Compound interest is related to loans and deposits made over longer periods. The interest is added to the initial sum at regular intervals, called conversion periods and the new account is used for calculating the interest for the next conversion period. The fraction of a year occupied by the conversation period is denoted by α so that the conversation period of one month is given by $\alpha=1/12$. Instead of saying that the conversation period is one month we say that the interest is compounded monthly.

【Example 1.1.1】 (Compound Interest)

An annual interest rate is $p\%$ and conversion period is α. Let $s(k)$ denote the amount on deposit after k conversion periods. Thus

---

❶ basic difference equations models 基本的差分方程模型

❷ compound interest 复利

❸ loan repayments 还款

$$s(1)=s(0)+s(0)\alpha p\%=s(0)(1+\alpha p\%)$$
$$s(k+1)=s(k)+s(k)\alpha p\%=s(k)(1+\alpha p\%)$$

Here, $s(k)$ follows the geometric progression so that

$$s(k)=(1+\alpha p\%)^k s(0) \tag{1.1.1}$$

gives the so-called compound interest formula.

If we want to measure time in years, the $k=t/\alpha$ where t is time in years. The (1.1.1) takes the form

$$s(t)=(1+\alpha p\%)^{t/\alpha} s(0) \tag{1.1.2}$$

If $s(0)=1$, then $s(1)=(1+\alpha p\%)^{1/\alpha}=1+p\%+\cdots>1+p\%$. So if the interest is compounded several times a year the increase in savings is bigger than if it was compounded annually.

Repayments are made at regular intervals and usually in equal amounts to reduce the loan and to pay the interest on the amount still owing.

【Example 1.1.2】 (Loan Repayments)

It is supposed that the compound interest at $p\%$ is charged on the outstanding debt with the conversion period equal to the same fraction α of the year as the period between the repayment. Let $D(0)$ be the initial debt to be repaid, for each k let $D(k)$ be the outstanding debt after k th repayment, and let the repayment made after each conversion period be R. Thus

$$D(k+1)=D(k)+\alpha p\% D(k)-R=D(k)(1+\alpha p\%)-R \tag{1.1.3}$$

We note that if the instalment was paid at the beginning of the conversion period, the equation would take a slightly different form

$$D(k+1)=D(k)-R+\alpha p\%(D(k)-R)=(D(k)-R)(1+\alpha p\%) \tag{1.1.4}$$

1.1.2 Some Money Related Models

【Example 1.1.3】 (The Cobweb Model[1])

We study the pricing of certain commodity. Let $S(n)$ be the number of units supplied in period n, $D(n)$ the number of units demanded in period n, and $p(n)$ the price per unit in period n. For simplicity, we assume that $D(n)$ depends only linearly on $p(n)$ and is denoted by

$$D(n)=-m_{\mathrm{d}} p(n)+b_{\mathrm{d}}, \quad m_{\mathrm{d}}>0, \quad b_{\mathrm{d}}>0 \tag{1.1.5}$$

This equation is referred to as the price-demanded curve. The constant m_{d} is the sensitivity of consumers to price. We also assume that the price-supply curve relates the supply in any period to the price one period before, i.e.

$$S(n)=m_{\mathrm{s}} p(n-1)+b_{\mathrm{s}}, \quad m_{\mathrm{s}}>0, \quad b_{\mathrm{s}}>0 \tag{1.1.6}$$

[1] Cobweb Model 蛛网模型

The constant m_s represents the sensitivity of suppliers to price. The slope of the demand curve is negative because an increase of one unit in price produces a decrease of m_d units in demand. Correspondingly, an increase of one unit in price causes an increase of m_s units in supply, creating a positive slope for that curve. A third assumption we make here is that the market price is the price at which the quantity demanded and the quantity supplied are equal, that is, at which $D(n)=S(n)$. Thus

$$-m_d p(n)+b_d=m_s p(n-1)+b_s$$

or

$$p(n)=Ap(n-1)+B=f(p(n-1)) \tag{1.1.7}$$

Where

$$A=-\frac{m_s}{m_d}, B=\frac{b_d-b_s}{m_d} \tag{1.1.8}$$

This equation is a first-order linear difference equation. The equilibrium price p^* is defined in economics as the price that results in an intersection of supply and demand curves. Also, p^* is the unique fixed point of $f(p)$ in (1.1.7), $p^*=B/(1-A)$.

【Example 1.1.4】 (National Income[1])

In market economy, the national income $Y(n)$ of a country in a given period n may be written as

$$Y(n)=C(n)+I(n)+G(n) \tag{1.1.9}$$

Where $C(n)$ is the consumer expenditure for purchase of consumer goods, $I(n)$ is the private investment for buying capital equipment, $G(n)$ is government expenditure, n is usually measured in years. There are various models for above functions. We now use widely accepted assumptions introduced by Samuelson.

The consumption satisfies

$$C(n)=\alpha Y(n-1), 0<\alpha<1$$

that is the consumer expenditure is proportional to the income in the preceding year. The investment satisfies

$$I(n)=\beta(C(n)-C(n-1)), \beta>0$$

so that the private investment is induced by the increase in consumption rather than that by the consumption itself. Finally, it is assumed that the government expenditure remains constant over the years and we let $G(n)=1$. So that (1.1.9) can be written as the second order linear difference equation:

$$Y(n+2)-\alpha(1+\beta)Y(n+1)+\alpha\beta Y(n)=1 \tag{1.1.10}$$

【Example 1.1.5】 (Gambler's Ruin[2])

---

[1] National Income 国民收入。由消费支出、投资、出口和进口几个主要因素组成

[2] Gambler's Ruin 赌徒破产

A gambler plays a sequence of games against an adversary. The probability that the gambler wins \$1 in any given game is q and the probability of his losing \$1 is $1-q$, where $0\leqslant q\leqslant 1$. He quits the game if he either loses all his money or wins a prescribed amount of N dollars. If the gambler runs out of money first, we say that he has been ruined. Let $p(n)$ denote the probability that the gambler will be ruined if he possesses n dollars. We build the difference equation satisfied by $p(n)$ using the following argument. He may be ruined in two ways. First, winning the next game; the probability of this event is q; then his fortune will be $n+1$, and the probability of being ruined is $p(n+1)$. Second, losing the next game; the probability of this event is $1-q$; then his fortune will be $n-1$, and the probability of being ruined is $p(n-1)$. Thus

$$p(n)=qp(n+1)+(1-q)p(n-1) \tag{1.1.11}$$

Replacing n by $n+1$, we get the second order linear difference equation

$$p(n+2)-\frac{1}{q}p(n+1)+\frac{1-q}{q}p(n)=0, n=0,1,\cdots,N \tag{1.1.12}$$

where $p(0)=1$and $p(N)=0$.

1.2 Difference equations of population theory

1.2.1 Single equations for unstructured population models

In many fields of human endeavor it is important to know how populations grow and what factors influence their growth.

【Example 1.2.1】 (Insect-type population)

Insects often have well-defined annual non-overlapping generation-adults lay eggs in spring/summer and then die. The eggs hatch into larvae which eat and grow and then overwinter in pupal stage. The adults emerge from the pupae in spring. We take the census of adults in the breeding seasons. It is natural to describe the population as the follow sequence $x(0)$, $x(1)$, $\cdots$, $x(n)$, $\cdots$where $x(n)$ is the number of adults in the n-th breeding season. We assume that there is a functional dependence between subsequent generations

$$x(n+1)=f(x(n)), n=0,1,2,\cdots \tag{1.2.1}$$

Let R_0be the average number of eggs laid by an adult. R_0 is called the basic reproductive ratio or the intrinsic growth rate. So that, the functional dependence (1.2.1) is

$$x(n+1)=R_0 x(n)=R_0{}^{n+1}x(0), n=0,1,2,\cdots \tag{1.2.2}$$

Which describe the situation that the size of the population is determined only by its fertility. The exponential equation (1. 2. 2) is also called *Malthusian equation*. If $R_0<1$, then $\lim_{n\to\infty} x(n)=0$, and population decreases towards extinction. But if $R_0>1$, then $x(n)$ increases indefinitely, and $\lim_{n\to\infty} x(n)=\infty$.

Such a behavior over long periods of time is not observed in any population so that we see that the model is over-simplified and requires corrections.

【Example 1. 2. 2】 (Models leading to nonlinear difference equations)

In a real population, some of the R_0 offspring produced by each adult will not survive to be counted as adults in the next census. If we denote $S(N)$ the survival rate, then the Malthusian equation is replaced by

$$N_{k+1}=R_0 S(N_k)N_k, k=0,1,2,\cdots \tag{1.2.3}$$

which may be alternatively written as

$$N_{k+1}=F(N_k)N_k, k=0,1,2,\cdots \tag{1.2.4}$$

where $F(N_k)$ is per capita production of a population of size N. Such models, with density dependent growth rate, lead to nonlinear equations.

We introduce most typical nonlinear models.

【Example 1. 2. 3】 (Beverton-Holt type models[1])

Let us look at the equation (1. 2. 4)

$$N_{k+1}=F(N_k)N_k, k=0,1,2,\cdots$$

where $F(N_k)=R_0S(N_K)$. We would like the model to display a *compensatory*[2] behavior; that is, mortality should balance the increase in numbers. For this we should have $NS(N)\approx$const. And for small N, $S(N)$ should be approximately 1 and the growth should be exponential with the growth rate R_0. A simple function of this form is $S(N)=\dfrac{1}{1+aN}$ leading to

$$N_{k+1}=\frac{R_0N_k}{1+aN_k}$$

If we introduce the concept of carrying capacity of the environment K and assume that if the population having reached K, it will stay K. That is, if $N_k=K$ for some k, then $N_{k+m}=K$ for all $m>0$, then

$$K=\frac{R_0K}{1+aK}$$

leading to $a=(R_0-1)/K$ and the resulting model, called the *Beverton-Holt model*, takes the form

---

[1] Beverton-Holt type models Beverton-Holt 模型，一类非常经典的离散时间种群模型，最早由 Beverton 和 Holt 于 1957 年提出来描述渔业的增长模型。

[2] compensatory 代偿，补偿性的

$$N_{k+1}=\frac{R_0 N_k}{1+\frac{R_0-1}{K}N_k} \tag{1.2.5}$$

A generalization of this model

$$N_{k+1}=\frac{R_0 N_k}{(1+aN_k)^b} \tag{1.2.6}$$

is called the *Hassell* or *again Beverton-Holt* model.

The Beverton-Holt models are best applied to semelparous insect populations but also used in the context of fisheries. For populations surviving to the next cycle it is more informative to write the difference equation in the follow form

$$N_{k+1}=N_k+R(N_k)N_k \tag{1.2.7}$$

so that the increase in the population is given by $R(N)=R_0 S(N)N$.

【Example 1.2.4】 (Logistic equation[1])

We assume that no adults die. The function R can have different forms but must satisfy the requirements:

(a) Due to overcrowding, $R(N)$ must decrease as N increases until N equals the carrying capacity K; then $R(K)=0$ and, as above, $N=K$ stops changing.

(b) Since for N much smaller than K there is small intra-species competition, we should observe an exponential growth of the population so that $R(N)\approx R_0$ as $N\to 0$; here R_0 is called the unrestricted growth rate of the population.

The constants R_0 and K are usually determined experimentally.

In the spirit of the mathematical modelling we start with the simplest function satisfying these requirements. The function R must be choose as

$$R(N)=-\frac{R_0}{K}N+R_0$$

Substituting this formula into (1.2.7) yields the discrete logistic equation

$$N_{k+1}=N_k+R_0 N_k\left(1-\frac{N_k}{K}\right) \tag{1.2.8}$$

which is still one of the most often used discrete equations of population dynamics.

1.2.2 Structured populations and linear systems of difference equations

The models introduced above have two problems. One is that all individuals in the population have the same age, the other is that the described population does not interact with each other. Trying to remedy these deficiencies leads to systems of equations.

---

[1] Logistic equation Logistic方程，又称虫口方程，是在研究种群增长问题中提出的一个简单模型，它具有非常复杂的动力学行为，为差分方程的复杂性的研究提供了一个范例。

We start with what possibly is the first formulated problem related to populations with age structure.

【Example 1. 2. 5】 (The Fibonacci Sequence[1])

This problem first appeared in 1202, in *liber abaci*, a book about the abacus, written by the famous Italian mathematician Leonardo di Pisa, better known as Fibonacci. The problem may be state as follows:

A certain man put a pair of rabbits in a place surrounded on all sides by a wall. How many rabbits can be produced from the pair of rabbits in a year if it is supposed that every month each pair begets a new pair which from the second month on becomes productive?

We assume that no death occur in the period of observation and the monthly census of population is taken just before births for this month taken place. Then, the number of pairs present in month $n+1$ is equal to the number pairs present in the month n, plus the number born in month n. Since rabbits become productive only two months after birth and produce only one pair per month, the number born in month n equals to the number present in the month $n-1$. Let $x(n)$ be the number of pairs at the end of month n, we obtain the so-called Fibonacci equation

$$x(n+1)=x(n)+x(n-1), \quad n=1,2,\cdots \tag{1.2.9}$$

This is a linear difference equation of second order.

We note that each month the population is represented by two classes of rabbits, adults and juveniles.

Let $x_1(n)$ be the number of adults and $x_2(n)$ be the number of juveniles in month n. Then the state of population in month n can be described by vector

$$\boldsymbol{v}(n)=\begin{bmatrix} x_1(n) \\ x_2(n) \end{bmatrix}$$

Since the number of juvenile (one-month old) pairs in month $n+1$ is equal to the number of adults in month n and the number of adults in month $n+1$ is the number of adults from month n and the number of juveniles from month n who became adults. In other words

$$x_1(n+1)=x_2(n)$$
$$x_2(n+1)=x_1(n)+x_2(n)$$

or, in a compact form

$$\boldsymbol{v}(n+1)=\begin{bmatrix} 0 & 1 \\ 1 & 1 \end{bmatrix}\boldsymbol{v}(n)=\boldsymbol{A}\boldsymbol{v}(n)=\boldsymbol{A}^n\boldsymbol{v}(0)$$

---

[1] Fibonacci Sequence　Fibonacci 数列，是在十分简单的假设下描述兔子数量增长的递推关系，是一种差分方程。

where

$$A=\begin{bmatrix}0 & 1\\ 1 & 1\end{bmatrix}$$

【Example 1.2.6】 (Models with age structure)

Assume that instead of pairs of individuals, we are tracking only females and that the census is taken immediately before the reproductive period. Further, assume that there is an oldest age class n and no individual can stay in an age class for more than one time period. Let s_i is the probability of survival from age $i-1$ to age i, and each female of age i produces m_i offspring in average. Hence, $s_1 m_i$ is the average number of female offspring produced by each female of age i who survived to census.

In this case, the evolution of the population can be described by the system of difference equations

$$\boldsymbol{v}(n+1)=\boldsymbol{A}\boldsymbol{v}(n)$$

where $\boldsymbol{A}$ is the $n\times n$ matrix

$$\boldsymbol{A}=\begin{bmatrix} s_1m_1 & s_1m_2 & \cdots & s_1m_{n-1} & s_1m_n\\ s_2 & 0 & \cdots & 0 & 0\\ 0 & s_3 & \cdots & 0 & 0\\ \cdots & \cdots & \cdots & \cdots & \cdots\\ 0 & 0 & \cdots & s_n & 0\end{bmatrix}$$

The matrix $\boldsymbol{A}$ is referred to as a Leslie matrix❶. To shorten notation we often denote $f_i=s_1m_i$, which is referred to as the age specific fertility. Assume that a fraction τ_i of i-population stays in the same population. This gives the matrix

$$\boldsymbol{A}=\begin{bmatrix} f_1+\tau_1 & f_2 & \cdots & f_{n-1} & f_n\\ s_2 & \tau_2 & \cdots & 0 & 0\\ 0 & s_3 & \cdots & 0 & 0\\ \cdots & \cdots & \cdots & \cdots & \cdots\\ 0 & 0 & \cdots & s_n & \tau_n\end{bmatrix}$$

which is a generalization of the Leslie matrix and called *Usher matrix*.

1.2.3 Markov chain❷

In 1906 the Russian mathematician A. A. Markov developed the concept of

---

❶ Leslie matrix　Leslie 矩阵，是科学家 Leslie 于 1945 年引进的一种数学方法，利用某一初始时刻种群的年龄结构现状，动态地预测种群的年龄结构和数量随时间的演变过程。

❷ Markov chain　马尔可夫链，是概率论和数理统计中具有马尔可夫性质且存在于离散的指数集和状态空间内的随机过程。它的命名来自俄国数学家安德雷·马尔可夫，以纪念其首次定义马尔可夫链和对其收敛性所做出的研究。

Markov chain.

【Example 1.2.7】 (Markov chain)

Suppose that we conduct some experiment with a set of n outcomes, or states, S_1, S_2, …, S_n. The experiment is repeated such that the probability (p_{ij}) of the state S_i, $1 \leqslant i \leqslant n$, occurring on the $(k+1)$th repetition depends only on the state S_j occurring on the kth repetition of the experiment. In other words, the system has no memory: the future state depends only on present state. In probability theory language, $p_{ij} = p(S_i \mid S_j)$ is the probability of S_i occurring on the next repetition, given that S_j occurred on the last repetition. Given that S_j has occurred in the last repetition, one of S_1, S_2, …, S_n must occur in the next repetition. Thus,

$$p_{1j} + p_{2j} + \cdots + p_{nj} = 1, 1 \leqslant j \leqslant n \tag{1.2.10}$$

Let $x_j(k)$ denote the probability that state S_j will occur on the k th repetition of the experiment, $1 \leqslant j \leqslant n$. Since one of S_1, S_2, …, S_n must occur on the k th repetition, it follows that

$$x_1(k) + x_2(k) + \cdots + x_n(k) = 1 \tag{1.2.11}$$

We obtain the k-dimensional system

$$\begin{aligned} x_1(k+1) &= p_{11}x_1(k) + p_{12}x_2(k) + \cdots + p_{1n}x_n(k) \\ x_2(k+1) &= p_{21}x_1(k) + p_{22}x_2(k) + \cdots + p_{2n}x_n(k) \\ &\cdots \\ x_n(k+1) &= p_{n1}x_1(k) + p_{n2}x_2(k) + \cdots + p_{nn}x_n(k) \end{aligned}$$

or

$$\boldsymbol{x}(k+1) = \boldsymbol{A}\boldsymbol{x}(k) \tag{1.2.12}$$

where $\boldsymbol{x}(k) = (x_1(k), x_2(k), \cdots, x_n(k))^{\mathrm{T}}$ is the probability vector and $\boldsymbol{A} = (p_{ij})_{n\times n}$ is a transition matrix.

The matrix $\mathbf{A}$ belongs to special class of matrices called *Markov matrices*. A matrix $\mathbf{A} = (a_{ij})_{n\times n}$ is said to be nonnegative (positive) if $a_{ij} \geqslant 0 (>0)$ for all entries a_{ij} of $\mathbf{A}$. A nonnegative $n \times n$ matrix $\mathbf{A}$ is said to be *Markov* if $\sum_{i=1}^{n} a_{ij} = 1$ for all $j = 1, 2, \cdots, n$.

A regular Markov chain is one in which $\boldsymbol{A}^m$ is positive for some positive integer m.

Chapter 2
Basic differential equations models[1]

As we observed in the previous section, the difference equation can be used to model quite a diverse phenomenon but their applicability is limited by the fact that the system should not change between subsequent time steps. These steps can vary from fraction of a second to years or centuries but they must stay fixed in the model. There are however numerous situations when the changes can occur instantaneously. These include growth of populations in which breeding is not restricted to specific seasons, motion of objects where the velocity and acceleration changes every instant, spread of epidemic with no restriction on infection times, and may others. In such cases we have to find the relations between the states of change of quantities relevant to the process. Rate of change are typically expressed as derivatives and thus continuous time modelling leads to differential equations that express relations between the derivatives than to difference equations that express relations between the states of the system in subsequent moments of time.

2.1 Equations related to financial mathematics

Many banks now advertise continuous compounding of interest which means that the conversion period tends to zero so that the interest is added to the account on the continual basis.

【Example 2.1.1】 Continuous compounded interest

An annual interest rate is p. Let $s(t)$ denote the amount on deposit. If we

[1] basic differential equations models 基本的微分方程模型

measure time in years, that is, Δt becomes the conversion period, then the increase in the deposit between time instants t and $t+\Delta t$ will be

$$s(t+\Delta t)=s(t)+s(t)\cdot\Delta t\cdot p \tag{2.1.1}$$

Dividing equation (2.1.1) by Δt and passing with Δt to zero, we have the differential equation

$$\frac{\mathrm{d}s}{\mathrm{d}t}=ps \tag{2.1.2}$$

This is a first order linear equation. It is easy to check that it has the solution

$$s(t)=s(0)\mathrm{e}^{pt} \tag{2.1.3}$$

where $s(0)$ is the initial deposit at time $t=0$.

Similar argument can be used for the loan repayment.

【Example 2.1.2】 Continuous loan repayment

Assume that the loan is being paid off continuously, a rate $\rho>0$ per annum. Then, after short period of time Δt the change in the debt D can be written, similarly to

$$D(t+\Delta t)=D(t)+\Delta t pD(t)-\rho\Delta t \tag{2.1.4}$$

As before, we divide by Δt and taking $\Delta t\rightarrow 0$ we obtain the following equation

$$\frac{\mathrm{d}D}{\mathrm{d}t}-pD=-\rho \tag{2.1.5}$$

with the initial condition $D(0)=D_0$ corresponding to the initial debt.

2.2 Continuous population models

In this subsection we will study first order differential equations which appear in the population growth theory. At first glance it appears that it is impossible to model the growth of species by differential equations since the population of any species always change by integer amounts. However if the population is very large and it is increased by a few in short time intervals, those changes are very small compared to the total population. Thus we make the approximation that large population changes continuously and differentiable in time.

Let $x(t)$ the number of members of a population. Large population if increase/decrease not too drastically than it is reasonable to approximate the function $x(t)$ by a continuous and differentiable function of t.

The problem we are going to solve here: if $x(t_0)$ is known, can be the number be predicted at subsequent times? In general the time scale depends on the population type: for humans in years, for bacteria in hours or years.

To model the growth we consider the rate at which the population changes. The instantaneous growth rate at time t is $\frac{dx}{dt}$. The instantaneous relative growth (or rate of change per individual) is

$$r(t)=\frac{1}{x}\frac{dx}{dt} \tag{2.2.1}$$

【Example 2.2.1】 Exponential growth (The Malthusian model[1])

Suppose the instantaneous relative growth is constant a, the equation (2.2.1) may be write as

$$\frac{dx}{dt}=ax \tag{2.2.2}$$

This model says that the rate of change of the population is proportional to the existing population. If $x(t_0)=x_0$, we obtained

$$x(t)=x_0 e^{a(t-t_0)} \tag{2.2.3}$$

If $a>0$ then this result gives an exponential growth.

If $a<0$ then this result gives an exponential decay and a is called growth (decay) rate.

A useful measure of the growth of a population is the doubling time. If $t_1-t_0=$time required for population to double, then at $x(t_1)=2x_0$ and so

$$2x_0=x_0 e^{a(t_1-t_0)}$$

Thus

$$t_1-t_0=\frac{\ln 2}{a} \tag{2.2.4}$$

【Example 2.2.2】 (Application to world human population)

In 1961 the world population was 3.06×10^9. Over the next 10 years it increased at relative growth rate of 2% per year. Calculate the doubling time.

Solution. Applying the formula, with $t_0=1961$, $a=0.02$, $x(t_0)=x_0=3.06\times 10^9$, we have

$$x(t)=3.06\times 10^9 e^{0.02(t-1961)}$$

and the doubling time is $\frac{\ln 2}{0.02}=34.7$ years. Estimates from 1700 to 1961 suggest that the doubling time was 35 years, which means that this model gives good agreement with reality.

What about future predictions: In 1991 a population of 5,575 million, in 2389 a population of 15,969,164 million (the surface area of the Earth is approximately

---

[1] Malthusian model 指数模型，又称马尔萨斯人口模型，是英国人口统计学家马尔萨斯（Malthusian）在 1798 年发表的《人口原理》一书中提出的。

16,700,500 million m^2).

That means that the exponential model gives unreasonable predictions for the future. For positive growth rate, when time tends to infinity, the population number tends to infinity which is in contradiction with the available resources or environment limitations. Therefore it is obvious that a new model has to be used which can give reasonable results for large times.

The exponential model (besides that it gives too large predictions for large times) does not reflect the competition between individuals, the limits of the resources and food supply. For many species with a finite food supply the population tends to an upper limit which depends on the environment conditions. The idea behind the logistic model says that while the population will continue to grow as time goes on, the rate at which it does this growing gets smaller.

【Example 2. 2. 3】 The logistic model (Verhulst❶-Pearl model)

Let us denote by c the reproductive parameter, $x(t)$ the population density and K the carrying capacity of the environment (or a limiting size of population). Then the proposed logistic (Verhulst 1844) is

$$\frac{\mathrm{d}x}{\mathrm{d}t}=cx\left(1-\frac{x}{K}\right) \tag{2.2.5}$$

In general, K is large. For small values of x, cx^2/K is negligible compared to cx and initially the population grows exponentially. When x is large, the cx^2/K term is no longer small and slows down the rate of increase of x.

The logistic equation can be interpreted if we expand out the logistic equation

$$\frac{\mathrm{d}x}{\mathrm{d}t}=cx-\frac{c}{K}x^2 \tag{2.2.6}$$

which suggests a situation when the birth rate is still constant c, but in which there is a mortality term, cx^2/K, when the population is high.

If $x>K$, then $\frac{\mathrm{d}x}{\mathrm{d}t}<0$ and the population decreases, when $x=K$ then the population growth is a constant.

It is easy to verify that

$$x=\frac{Kx_0}{x_0+(K-x_0)\mathrm{e}^{-ct}} \tag{2.2.7}$$

Since $\mathrm{e}^{-ct}\to 0$ as $t\to\infty$ for $c>0$ it follows that the above form of the solution that $x\to K$ as $t\to\infty$ for any x_0.

The equation (2.2.5) has two population levels $x=0$ and $x=K$ for which $\frac{\mathrm{d}x}{\mathrm{d}t}=0$. These

---

❶ Verhulst 费尔哈斯特。荷兰生物学家（1804—1849）。

are called equilibrium levels[1]. Suppose that the population is closed to one of these levels at some time and will remain dose to that level for all t, we say that it is locally stable[2], if it does not remain close then it is unstable.

The point $x=K$ is locally stable but $x=0$ is unstable. We can gain the information as follows:

For $x=K$ write $x=K+X(t)$, where $X(t)$ is a small quantity; then $\frac{\mathrm{d}x}{\mathrm{d}t}=\frac{\mathrm{d}X}{\mathrm{d}t}$, since K is a constant, and from the logistic equation

$$\frac{\mathrm{d}X}{\mathrm{d}t}=c(K+X)\left(1-\frac{K+X}{K}\right)=-cX-\frac{cX^2}{K} \tag{2.2.8}$$

Since $X(t)$ is a small quantity we can neglect the term containing the X^2 term, so

$$\frac{\mathrm{d}X}{\mathrm{d}t}=-cX \tag{2.2.9}$$

and

$$X=A\mathrm{e}^{-ct} \tag{2.2.10}$$

Since $c>0$, $x\to 0$ as $t\to\infty$, i.e. x remains close to $x=K$ and so is locally stable.

【Example 2.2.4】 Harvesting

The population of fish in the North Sea can be described by the logistic equation

$$\frac{\mathrm{d}x}{\mathrm{d}t}=cx\left(1-\frac{x}{K}\right)=cx-sx^2 \tag{2.2.11}$$

where $s=c/K$ is a positive number.

Supposed that the fish are caught (harvested) at a constant rate h, i.e. in absence of all other factors

$$\frac{\mathrm{d}x}{\mathrm{d}t}=-h \tag{2.2.12}$$

To account for harvesting the logistic equation must be modified to

$$\frac{\mathrm{d}x}{\mathrm{d}t}=cx-sx^2-h\ . \tag{2.2.13}$$

Let us first discuss the equilibrium levels of this model. The equilibrium levels are at $x=x_\mathrm{e}$, such that $\frac{\mathrm{d}x}{\mathrm{d}t}=0$, i.e.

$$cx_\mathrm{e}-sx_\mathrm{e}^2-h=0 \tag{2.2.14}$$

or

$$x_\mathrm{e}^2-Kx_\mathrm{e}+\frac{h}{s}=0 \tag{2.2.15}$$

---

[1] equilibrium levels 平衡水平

[2] locally stable 局部稳定

Solving this second order equation we obtain

$$x_e=\frac{K\pm\sqrt{K^2-4\frac{h}{s}}}{2}=\frac{K}{2}\pm\sqrt{\frac{K^2}{4}-\frac{h}{s}} \qquad (2.2.16)$$

Case (a) If the determinant is positive, i.e. $\frac{K^2}{4}-\frac{h}{s}>0$, then $h<\frac{sK^2}{4}$ (moderate fishing).

In this case, there are two positive real roots (i.e. two equilibrium levels) x_1 and x_2 where

$$x_1=\frac{K}{2}+\sqrt{\frac{K^2}{4}-\frac{h}{s}}<K$$

$$x_2=\frac{K}{2}-\sqrt{\frac{K^2}{4}-\frac{h}{s}}$$

$$x_2<x_1<K$$

and

$$\frac{dx}{dt}=cx-sx^2-h=-s(x-x_1)(x-x_2) \qquad (2.2.17)$$

Case (b) If the determinant is negative, i.e. $\frac{K^2}{4}-\frac{h}{s}<0$, then $h>\frac{sK^2}{4}$ (over intensive fishing).

In this case there are no real roots, i.e. there are no equilibrium levels. Completing the squares, dx/dt can be written as

$$\frac{dx}{dt}=-s\left(x^2-Kx+\frac{h}{s}\right)=-s\left[\left(x-\frac{K}{2}\right)^2-\frac{K^2}{4}+\frac{h}{s}\right]<0. \qquad (2.2.18)$$

Since dx/dt never reaches 0, we have no equilibrium levels.

Case (c) If the determinant is zero, i.e. $\frac{K^2}{4}-\frac{h}{s}=0$. Equation (2.2.15) has a repeated root $x_e=\frac{K}{2}$.

From equation (2.2.13)

$$\frac{dx}{dt}=-s\left(x-\frac{K}{2}\right)^2<0 \qquad (2.2.19)$$

for all values of x except $x=x_e$.

2.3 Equations of motion: second order equations

Second order differential equations appear often as equations of motion. The motion of a particle along a straight line (the x-axis) is described by its position function

$$x=f(t) \tag{2.3.1}$$

Gives its x-coordinate at time t. The velocity of the particle is defined to be

$$v(t)=f'(t) \tag{2.3.2}$$

that is,

$$v=\frac{dx}{dt} \tag{2.3.2'}$$

The acceleration $a(t)$ is $a(t)=v'(t)=x''(t)$; in Leibniz notation,

$$a=\frac{dv}{dt}=\frac{d^2x}{dt^2} \tag{2.3.3}$$

Newton's second law of motion says that if a force $F(t)$ acts on the particle and is directed along its line of motion, then

$$ma(t)=F(t) \tag{2.3.4}$$

that is

$$F=ma \tag{2.3.5}$$

where m is the mass of the particle.

【Example 2.3.1】 The problem of forced harmonic oscillator[1]

In order to understand the problem of forced harmonic motion, first let us see what equation will describe the free harmonic oscillator. The typical harmonic oscillator is the mass-on-spring system, which is described the equation

$$m\frac{d^2x}{dt^2}+b\frac{dx}{dt}+kx=0 \tag{2.3.6}$$

where m is the mass, k is the spring constant, and b is the coefficient of the "viscous" damping term, which represents a force proportional to the speed of the mass (of course, x represents the displacement of the mass from its equilibrium point). In an ideal environment, the damping term can be set to zero, so the governing equation will be

$$m\frac{d^2x}{dt^2}+kx=0 \tag{2.3.7}$$

The solution of equation (2.3.7) has the simple sinusoidal form

$$x(t)=A\cos(\omega_0 t+\varphi) \tag{2.3.8}$$

where $\omega_0=\sqrt{m/k}$ is the frequency of free oscillations and the constants A and φ can be determined once initial conditions are known. The constant ω_0 is called the natural frequency of the system since it is the frequency the system will oscillate when left alone.

Now let us return to the case of a damped oscillator and write the equation in the form

---

[1] harmonic oscillator 谐振动。如弹簧的振动为谐振动。

$$\frac{d^2x}{dt^2}+\frac{b}{m}\frac{dx}{dt}+\frac{k}{m}x=0 \tag{2.3.9}$$

By substituting ω_0^2 for m/k and $2p$ for b/k we have

$$\frac{d^2x}{dt^2}+2p\frac{dx}{dt}+\omega_0^2x=0 \tag{2.3.10}$$

The solution of equation (2.3.10) has the form

$$x(t)=Ae^{s_1t}+Be^{s_2t} \tag{2.3.11}$$

where $s_{1,2}=-p\pm\sqrt{p^2-\omega_0{}^2}$. But the quality varies with the relative magnitudes of p and ω_0 (we only interested in case when $p>0$). According to the values of $s_{1,2}$, we can distinguish three cases.

Case (a) $p<\omega_0$

In this case, the inside of the radical is negative. Thus, we can find that

$$x(t)=A_de^{-pt}\cos(\omega_dt+\varphi_d) \tag{2.3.12}$$

where $\omega_d=\sqrt{\omega_0^2-p^2}$.

Once again, the constant are determined by the initial conditions. This motion is still sinusoidal (oscillatory) but its envelope is exponential. The angular frequency of the motion is now modified from ω_0 to ω_d and $\omega_d<\omega_0$. This kind of oscillatory motion is called underdamped oscillatory motion[1].

Case (b) $p>\omega_0$

Here, the inside of the radical is positive, so we have two distinct, real roots. The form of the solution then becomes

$$x(t)=A_1e^{\lambda_1t}+A_2e^{\lambda_2t} \tag{2.3.13}$$

where $\lambda_1=-p+\sqrt{p^2-\omega_0{}^2}$ and $\lambda_2=-p-\sqrt{p^2-\omega_0{}^2}$.

In a similar manner, the constant A_1 and A_2 are determined by the initial conditions. Since all exponents are real, it is obvious that there will be no oscillations at all and the motion dies off exponentially. This case is also known as the overdamped case.

Case (c) $p=\omega_0$

Now the quantities under the radical are all zero and the form of the solution becomes

$$x(t)=(A_c+B_ct)e^{-pt} \tag{2.3.14}$$

The oscillations die off very quickly and there will be no oscillation. This case is also known as the critically damped case.

With all these cases now being clear, we can turn our attention to the case of a forced oscillation where the forcing term (driver) is sinusoidal, i. e.

---

[1] underdamped oscillatory motion 欠阻尼自由振动

$$m\frac{d^2x}{dt^2}+b\frac{dx}{dt}+kx=F\cos\omega t \qquad (2.3.15)$$

Here, F is the maximum force, and ω is the driving frequency. Dividing through by m again, we obtain

$$\frac{d^2x}{dt^2}+\frac{b}{m}\frac{dx}{dt}+\frac{k}{m}x=\frac{F}{m}\cos\omega t \qquad (2.3.16)$$

This is a second order linear differential equation.

【Example 2.3.2】 Motion in a changing gravitational field

According to Newton's law of gravitation, the gravitational force of attraction between two point masses m and M located at a distance d apart is given by

$$F=G\frac{mM}{d^2} \qquad (2.3.17)$$

where G is the gravitational constant ($G\approx 6.6726\times 10^{-11}\,\mathrm{N}\cdot(\mathrm{m/kg})^2$ in mks units). The formula is also valid if either or both of the two masses are homogeneous spheres; in this case, d is the distance between the objects' centers. Since the Earth's surface the force is equal to $F=mg$, the gravitational force exerted on a body of mass m at a distance y above the surface is given by

$$F=-\frac{mgR^2}{(y+R)^2} \qquad (2.3.18)$$

where the minus sign indicates that the force acts towards Earth's center. Thus the equation of motion of an object of mass m projected upward from the surface is

$$m\frac{d^2y}{dt^2}=-\frac{mgR^2}{(y+R)^2}-c\left(\frac{dy}{dt}\right)^2 \qquad (2.3.19)$$

where the last term represents the air resistance. This is a second order nonlinear differential equation.

2.4 Modelling interacting quantities systems of differential equations

Two species model[1]

Suppose an island supports a crop of foliage and populations of foxes and rabbits. The foxes eat the rabbits. The rabbits eat the foliage (i.e. limited supply). What happens with the populations?

Assuming no other interference with the system, we might expect the two populations to oscillate.

[1] two species model 两生物种群模型

【Example 2.4.1】 Lotka-Volterra[1] model (Predator-prey)[2]

Let $x(t)$ be the population of rabbits at time t. With no foxes around, we assume that $x(t)$ obeys the logistic equation

$$\frac{dx}{dt}=ax-bx^2,\ a>0, b>0 \tag{2.4.1}$$

and a/b being the carrying capacity or saturation level.

Let $y(t)$ be the population of foxes at time t. Assuming that the number of encounters per unit time between foxes and rabbits is proportional to x and y and that a certain proportion of these results in rabbit being eaten, i.e. the rate of decrease in the rabbit population is proportional to xy. In this case the differential equation becomes

$$\frac{dx}{dt}=ax-bx^2-cxy,\ a, b, c>0 \tag{2.4.2}$$

If no rabbits are present, we assume foxes die out exponentially, i.e.

$$\frac{dy}{dt}=-py,\ p>0 \tag{2.4.3}$$

With rabbits present, the rate of increase of foxes is proportional to the number of successful encounters with rabbits and so the differential equation becomes

$$\frac{dy}{dt}=-py+qxy,\ p, q>0 \tag{2.4.4}$$

The two populations are therefore determined by the coupled pair of nonlinear ordinary differential equations (2.4.2) and (2.4.4). Putting $b=0$ gives the Lotka-Volterra equations

$$\begin{cases}\dfrac{dx}{dt}=ax-cxy\\[2ex]\dfrac{dy}{dt}=-py+qxy\end{cases} \tag{2.4.5}$$

This is a two-dimensional nonlinear system of first order differential equations.

【Example 2.4.2】 Competing model[3]

Two species of warbler compete for the same kind of insect in a spruce forest. The equations governing their population, x and y, in appropriate units, are

$$\begin{cases}\dfrac{dx}{dt}=ax-cxy\\[2ex]\dfrac{dy}{dt}=py-qxy\end{cases} \tag{2.4.6}$$

---

[1] Volterra，沃特拉。意大利数学家（1860—1940）。

[2] predator-prey 捕食-被捕食

[3] competing model 竞争模型

where a, c, p, $q>0$.

This system of equations is called the competing model. If the constants c and q are positive numbers, the system of equations is called the *symbiotic model*[1].

【Example 2. 4. 3】 Lorenz equations[2]

The famous Lorenz system of differential equations is a simplified system describing the motion of atmospheric flow. It is established by the mathematical meteorologist E. N. Lorenz, who later described his discovery as follows.

By the middle 1950's "numerical weather prediction;" i. e. , forecasting by numerically integrating such approximations to the atmospheric equations as could feasibly be handled, was very much in vogue, despite the rather mediocre results which it was then yielding. A smaller but determined group favored statistical prediction. …I was skeptical, and decided to test the idea by applying the statistical method to a set of artificial data, generated by solving a system of equations numerically…. The first task was to find a suitable system of equations to solve . …The system would have to be simple enough… and the general solution would have to be aperiodic, since the statistical prediction of a periodic series would have to be a trivial matter, once the periodicity had been detected… [In the course of talks with Dr. Barry Saltzman] he showed me some work on thermal convections, in which he used a system of seven ordinary differential equations. Most of his solutions soon acquired periodic behavior, but one solution refused to settle down. Moreover, in this solution four of the variables appeared to approach zero. Presumably the equations governing the remaining three variables, with the terms containing the four variables eliminated…. [Quoted in E. Hairer, S. P. Norsett, and G. Solving Ordinary Differential Equations (New York: Springer-Verlag, 1987).]

The system of Lorenz equations is given by

$$\begin{cases} \dfrac{\mathrm{d}x}{\mathrm{d}t}=a(y-x) \\ \dfrac{\mathrm{d}y}{\mathrm{d}t}=-xz+cx-y \\ \dfrac{\mathrm{d}z}{\mathrm{d}t}=xy-bz \end{cases} \tag{2.4.7}$$

Where the three constants $a=10$, $b=\dfrac{8}{3}$ and $c=28$.

This is a nonlinear three-dimensional system of first order differential equations.

---

[1] symbiotic model 共生模型，两个种群互相促进、互为依赖

[2] Lorenz，洛伦茨。美国气象学家（1853—1928）。洛伦茨方程是其 1963 年发表在美国气象学报的描述大气对流现象的一个简化方程组. 洛伦茨方程被称为混沌现象的第一例。

Chapter 3

Solution and applications of difference equations

3.1 Linear first-order difference equations[1]

In this section we study the simplest special cases of first-order difference equation, namely, linear equation.

Difference equations usually describe the evolution of certain phenomena over the course of time.

If a certain population has discrete generations, the size of the $(n+1)$st generation $x(n+1)$ is a function of the nth generation $x(n)$. This relation expresses itself in the difference equation

$$x(n+1)=f(x(n)) \tag{3.1.1}$$

We may look at this problem from another point of view. Starting from a point x_0, one may generate the sequence

$$x_0, f(x_0), f(f(x_0)), f(f(f(x_0))), \cdots$$

For convenience we adopt the notation

$$f^0(x_0)=x_0, f^1(x_0)=f(x_0), f^2(x_0)=f(f(x_0)), f^3(x_0)=f(f(f(x_0))), \cdots$$

$f^1(x_0)$ is called the first iterate[2] of x_0 under f; $f^2(x_0)$ is called the second iterate of x_0 under f; more generally, $f^n(x_0)$ is called the nth iterate of x_0 under f. The set of all positive iterates $\{f^n(x_0): n \geqslant 0\}$ is called the positive orbit[3] of x_0 and will denoted by $O(x_0)$. This iterative procedure is an example of a discrete dy-

---

[1] linear first-order difference equations 一阶线性差分方程

[2] iterate 迭代

[3] positive orbit 轨道

namical system[1]. Let $x(n)=f^n(x_0)$, we have

$$x(n+1)=f^{n+1}(x_0)=f(f^n(x_0))=f(x(n))$$

and hence we obtain equation (3.1.1).

【Example 3.1.1】 Let $x(n+1)=x^2(n)$, $x(0)=0.6$. Computing $x(n)$ recursively, we have

$$x(1)=(0.6)^2=0.36$$
$$x(2)=(0.6)^4=0.1296$$
$$x(3)=(0.6)^8=0.01679676$$
$$\cdots$$
$$x(n)=(0.6)^{2n}$$

Definition 3.1.1. If f is a function of two variables, that is $f: N_0\times R\rightarrow R$, where $N_0=\{0, 1, 2, \cdots\}$ is the set of natural numbers enlarged by 0 and R is the set of real numbers,

$$x(n+1)=f(n,x(n)),x(0)=x_0 \tag{3.1.2}$$

then the equation (3.1.2) is called first-order non-autonomous difference equation[2], and $x(0)=x_0$ is called initial condition. The equation (3.1.1) is called first-order autonomous difference equation[3].

The form

$$x(n+1)=a(n)x(n)+g(n),x(0)=x_0,n=0,1,2,\cdots \tag{3.1.3}$$

where $\{a(n)\}$ and $\{g(n)\}$ are given sequences, is a typical linear nonhomogeneous first-order difference equation. If $g(n)=0$, then (3.1.3) is a linear nonhomogeneous first-order difference equation[4]. Given the constants A and B a difference equation of the form

$$x(n+1)=Ax(n)+B,x(0)=x_0,n=0,1,2,\cdots \tag{3.1.4}$$

is a special case of first order difference equation (3.1.3) and called first-order linear difference equation. In order to solve the equation (3.1.4), we write

$$\begin{aligned}x(n)&=Ax(n-1)+B=A[Ax(n-2)+B]+B\\&=A^2x(n-2)+B(A+1)=A^2[Ax(n-3)+B]+B(A+1)\\&=A^3x(n-3)+B(A^2+A+1)\\&\cdots\\&=A^nx_0+B(A^{n-1}+A^{n-2}+\cdots+A+1)\end{aligned} \tag{3.1.5}$$

If $A=1$, then

---

[1] discrete dynamical system 离散动力系统

[2] first-order non-autonomous difference equation 一阶非自治差分方程

[3] first-order autonomous difference equation 一阶自治差分方程

[4] linear nonhomogeneous first-order difference equation 一阶线性非齐次差分方程

$$x(n)=x_0+nB, n=0,1,2,\cdots$$

is the solution of the difference equation $x(n+1)=x(n)+B$.

If $A\neq 1$, then

$$A^{n-1}+A^{n-2}+\cdots+A+1=\frac{1-A^n}{1-A}$$

Hence,

$$x(n)=A^n x_0+B\frac{1-A^n}{1-A}, n=0,1,2,\cdots \tag{3.1.6}$$

is the solution of the first-order linear difference equation (3.1.4), when $A\neq 1$.

The equation

$$x(n+1)=Ax(n)+g(n),\quad x(0)=x_0, n=0,1,2,\cdots \tag{3.1.7}$$

is also a special case of first order difference equation (3.1.3), which appears in many applications. By simple iterations, we obtain that

$$\begin{aligned}
x(n) &= Ax(n-1)+g(n-1)=A(Ax(n-2)+g(n-2))+g(n-1)\\
&= A^2x(n-2)+Ag(n-2)+g(n-1)\\
&= A^2(Ax(n-3)+g(n-3))+Ag(n-2)+g(n-1)\\
&= A^3x(n-3)+(A^2g(n-3)+Ag(n-2)+g(n-1))\\
&\cdots\\
&= A^n x_0+\sum_{k=0}^{n-1}A^{n-k-1}g(k)
\end{aligned} \tag{3.1.8}$$

It is easy to see that

$$x(n)=A^n x_0+\sum_{k=0}^{n-1}A^{n-k-1}g(k) \tag{3.1.9}$$

is the solution of equation (3.1.7).

The unique solution of the nonhomogeneous (3.1.3) is

$$x(n)=x_0\prod_{i=0}^{n-1}a(i)+\sum_{k=0}^{n-1}\left(\prod_{i=k+1}^{n-1}a(i)\right)g(k) \tag{3.1.10}$$

It is easy to show that by using mathematical induction[1].

【Example 3.1.2】 Find a solution for the equation

$$x(n+1)=\frac{1}{2}x(n)+\frac{1}{2}, x(0)=\frac{1}{2}$$

Solution. From equation (3.1.6), we have

$$x(n)=\left(\frac{1}{2}\right)^n\frac{1}{2}+\frac{1}{2}\frac{1-\left(\frac{1}{2}\right)^n}{1-\frac{1}{2}}=1-\left(\frac{1}{2}\right)^{n+1}$$

【Example 3.1.3】 Find a solution for the equation

---

[1] mathematical induction 数学归纳法

$$x(n+1)=3x(n)+2^n, \quad x(0)=\frac{1}{3}$$

Solution. From equation (3.1.9), we have

$$x(n)=3^n \cdot \frac{1}{3}+\sum_{k=0}^{n-1} 3^{n-k-1} 2^k=3^{n-1}+3^{n-1} \sum_{k=0}^{n-1}\left(\frac{2}{3}\right)^k$$

$$=3^{n-1}+3^{n-1} \frac{1-\left(\frac{2}{3}\right)^n}{1-\frac{2}{3}}$$

$$=3^{n-1}+3^n-2^n.$$

【Example 3.1.4】 A drug is administered once every four hours. Let $x(n)$ be the amount of the drug in blood system at nth interval. The body eliminates a certain fraction p of the drug during each time interval. If the amount administered is x_0, find $x(n)$ and $\lim_{n \to \infty} x(n)$.

Solution. Since the amount of drug in blood system at time $n+1$ is equal to the amount at time n minus the fraction p that has been eliminated from the body, plus the new dosage x_0, we arrive at the following equation:

$$x(n+1)=(1-p)x(n)+x_0$$

Using equation (3.1.6), we find the solution of the above equation

$$x(n)=(1-p)^n x_0+x_0 \frac{1-(1-p)^n}{p}$$

$$=(1-\frac{1}{p})(1-p)^n x_0+\frac{x_0}{p}$$

Hence,

$$\lim_{n \to \infty} x(n)=\frac{x_0}{p}.$$

3.2 Difference calculus and general theory of linear difference equations

In this section we will study linear difference equations of high order, namely, those involving a single dependent variable. Such equations arise in almost every field of scientific inquiry, from population dynamics (the study of a single species) to economics (the study of a single commodity) to physics (the study of a single body). We start this section by introduce some rudiments of difference calculus that are essential in the study of linear equations.

3.2.1 Difference calculus

Difference calculus is the discrete analogue of the familiar differential and integral calculus. In this section we introduce some very basic properties of difference operator[1] and shift operator[2].

Definition 3.2.1. If $f(x)$ is a function from $R \to R$, then $\Delta f(x) = f(x+1) - f(x)$ is said to be the difference of $f(x)$ at x, and Δ is the difference operator. $Ef(x) = f(x+1)$ is said to be the shift of $f(x)$ at x and E is the shift operator. Let I be the identity operator[3], i.e. $If(x) = f(x)$.

It is easy to show that both difference operator and shift operator are linear operators, and satisfy the following two equations $\Delta = E - I$, $E = \Delta + I$. Hence,

$$\Delta^n = (E - I)^n = \sum_{k=0}^{n} (-1)^k C_n^k E^{n-k} \tag{3.2.1}$$

$$E^n = (\Delta + I)^n = \sum_{k=0}^{n} C_n^k \Delta^{n-k} \tag{3.2.2}$$

Where $C_n^k = \dfrac{n!}{k!\,(n-k)!}$, and $\Delta^0 = E^0 = I$. Similarly, one may show that

$$f(x+n) = E^n f(x) = \sum_{k=0}^{n} C_n^k \Delta^{n-k} f(x) \tag{3.2.3}$$

Lemma 3.2.1. The following statements hold:

(1) $\sum\limits_{k=n_0}^{n-1} \Delta f(k) = f(n) - f(n_0)$,

(2) $\Delta\left(\sum\limits_{k=n_0}^{n-1} f(k)\right) = f(n) - f(n_0)$.

The proof remains as Exercises 3.7, Problem 5.

【Example 3.2.1】 Let $f(x) = x^2 + 2x - 1$, $g(x) = 2^x$, $h(k) = 2k - 5 \cdot 3^k + k^2 + 4$, calculate

$\Delta f(x)$, $\Delta^2 f(x)$, $\Delta g(x)$, $\Delta^2 g(x)$, $\Delta h(k)$, $\Delta^2 h(k)$.

Solution. Using the definition of Δ, we obtain

$$\Delta f(x) = f(x+1) - f(x) = [(x+1)^2 + 2(x+1) - 1] - [x^2 + 2x - 1] = 2x + 3$$

$$\Delta^2 f(x) = \Delta f(x+1) - \Delta f(x) = [2(x+1) + 3] - [2x + 3] = 2$$

$$\Delta g(x) = g(x+1) - g(x) = 2^{x+1} - 2^x = 2^x$$

$$\Delta^2 g(x) = \Delta g(x+1) - \Delta g(x) = 2^{(x+1)} - 2^x = 2^x$$

---

[1] difference operator 差分算子，与微积分中导数的定义类似，也具有和导数类似的线性性质等

[2] shift operator 位移算子，也称移位算子

[3] identity operator 恒等算子

$$\Delta h(k)=h(k+1)-h(k)$$
$$=[2^{k+1}-5\cdot 3^{k+1}+(k+1)^2+4]-(2^k-5\cdot 3^k+k^2+4)$$
$$=2^k-10\cdot 3^k+2k+1$$
$$\Delta^2 h(k)=\Delta h(k+1)-\Delta h(k)$$
$$=[2^{k+1}-10\cdot 3^{k+1}+2(k+1)+1]-(2^k-10\cdot 3^k+2k+1)$$
$$=2^k-20\cdot 3^k+2$$

【Example 3. 2. 2】 Let $P(n)=a_0n^k+a_1n^{k-1}+\cdots+a_k$ be a polynomial of degree k❶ prove that $\Delta^{k+i}P(n)=0$ for $i\in Z^+$.

Solution.

$$\Delta P(n)=P(n+1)-P(n)$$
$$=[a_0(n-1)^k+a_1(n-1)^{k-1}+\cdots+a_k]-(a_0n^k+a_1n^{k-1}+\cdots+a_k)$$
$$=a_0kn^{k-1}+P_1(n)$$

$P_1(n)$ is a polynomial of degree lower than $k-1$. So that $\Delta P(n)$ is a polynomial of degree $k-1$.

Similarly, we can show that

$$\Delta^2 P(n)=\Delta P(n+1)-\Delta P(n)$$
$$=[a_0k(n+1)^{k-1}+P_1(n+1)]-[a_0kn^{k-1}+P_1(n)]$$
$$=a_0k(k-1)n^{k-2}+P_2(n)$$

where $P_2(n)$ is a polynomial of degree lower than $k-2$ and $\Delta^2 P(n)$ is a polynomial of degree $k-2$.

Carrying this process k times, we obtains

$$\Delta^k P(n)=a_0k! \quad (3.2.4)$$

Thus,

$$\Delta^{k+i}P(n)=0 \quad \text{for } i\in Z^+ \quad (3.2.5)$$

【Example 3. 2. 3】 If $f(x)$ is a polynomial of degree three, and $f(0)=-4$, $f(1)=-6$, $f(2)=-8$. Find the value of $f(4)$.

Solution. Using the formulas (3. 2. 1) (3. 2. 2) (3. 2. 5), we obtain

$$f(4)=E^4f(0)=(\Delta+I)^4f(0)=\Delta^4f(0)+4\Delta^3f(0)+6\Delta^2f(0)+f(0)$$
$$\Delta f(0)=f(1)-f(0)=-2$$
$$\Delta^2f(0)=(E-I)^2f(0)=E^2f(0)-2Ef(0)+f(0)=f(2)-2f(1)+f(0)=0$$
$$\Delta^3f(0)=(E-I)^3f(0)=f(3)-3f(2)+3f(1)-f(0)=6$$
$$\Delta^4f(0)=0$$

Thus,

$$f(4)=\Delta^4f(0)+4\Delta^3f(0)+6\Delta^2f(0)+f(0)=24-8-4=12$$

We can also find the value of $f(4)$ by the method of undetermined coefficients.

---

❶ polynomial of degree k k 次多项式

We now discuss the action of a polynomial of degree k in the shift operator E on the term b^n, for any constant b. Let $P(E)=a_0E^k+a_1E^{k-1}+\cdots+a_kI$ be a polynomial of degree k in E. Then

$$\begin{aligned} P(E)b^n &= a_0b^{n+k}+a_1b^{n+k-1}+\cdots+a_kb^n \\ &= (a_0b^k+a_1b^{k-1}+\cdots+a_k)b^n \\ &= P(b)b^n \end{aligned}$$

Lemma 3.2.2. Let $P(E)=a_0E^k+a_1E^{k-1}+\cdots+a_kI$ and $g(n)$ be a discrete function. Then

$$P(E)(b^ng(n))=b^nP(bE)g(n)$$

The proof is left to the readers as Exercises 3.7, Problem 6.

3.2.2 General theory of linear difference equations

Definition 3.2.2. Let $p_i(n)$ and $g(n)$ be real-valued functions defined for $n\geqslant 0$ and $p_k(n)\neq 0$ for all $n\geqslant 0$. Then the form

$$y(n+k)+p_1(n)y(n+k-1)+\cdots+p_k(n)y(n)=g(n) \qquad (3.2.6)$$

is said to be a kth-order nonhomogeneous linear difference equation[1]. If $g(n)$ is identically zero, then

$$y(n+k)+p_1(n)y(n+k-1)+\cdots+p_k(n)y(n)=0 \qquad (3.2.7)$$

is said to be a homogeneous equation[2].

Definition 3.2.3. A sequence $\{y(n)\}_0^\infty$ or simply $y(n)$ is said to be a solution of (3.2.6) if it satisfies the equation.

Equation (3.2.6) may be written in the form

$$y(n+k)=-p_1(n)y(n+k-1)-\cdots-p_k(n)y(n)+g(n) \qquad (3.2.8)$$

By letting $n=0$ in equation (3.2.8), we obtain $y(k)$ in terms of $y(k-1)$, $y(k-2)$, $\cdots$, $y(0)$. Explicitly, we have

$$y(k)=-p_1(0)y(k-1)-\cdots-p_k(0)y(0)+g(0)$$

By letting $n=1$in equation (3.2.8), we have

$$y(k+1)=-p_1(1)y(k)-\cdots-p_k(1)y(1)+g(1)$$

Once $y(k)$ is computed, we can evaluate $y(k+1)$. By repeating the above process, it is possible to evaluate all $y(n)$ for $n\geqslant k$. We will illustrate the above procedure by an example.

【Example 3.2.4】 Consider the following difference equation

$$y(n+3)-\frac{n}{n+1}y(n+2)+ny(n+1)-3y(n)=n \qquad (3.2.9)$$

where $y(1)=0$, $y(2)=-1$, and $y(3)=1$. Find the values of $y(4)$, $y(5)$, $y(6)$,

---

[1] kth-order nonhomogeneous linear difference equation k 阶线性非齐次差分方程

[2] homogeneous equation 齐次方程

$y(7)$.

Solution. First we write the third-order difference equation in the convenient form

$$y(n+3)=\frac{n}{n+1}y(n+2)-ny(n+1)+3y(n)+n \tag{3.2.10}$$

Letting $n=1$ in equation (3.2.10), we have

$$y(4)=\frac{1}{2}y(3)-y(2)+3y(1)+1=5$$

For $n=2$,

$$y(5)=\frac{2}{3}y(4)-2y(3)+3y(2)+2=-\frac{4}{3}$$

For $n=3$,

$$y(6)=\frac{3}{4}y(5)-3y(4)+3y(3)+3=-\frac{5}{2}$$

For $n=4$,

$$y(7)=\frac{4}{5}y(6)-4y(5)+3y(4)+4=14\frac{5}{6}$$

If we specify the initial data of the equation (3.2.6), we lead to the corresponding *initial value problem*[1]

$$\begin{cases} y(n+k)+p_1(n)y(n+k-1)+\cdots+p_k(n)y(n)=g(n) \\ y(0)=a_0, y(1)=a_1, \cdots, y(k-1)=a_{k-1} \end{cases} \tag{3.2.11}$$

where a_0, a_1, $\cdots$, a_{k-1}, are all real numbers.

It is easy to obtain the following result.

Theorem 3.2.1. The initial value problem (3.2.11) has a unique solution $y(n)$.

We are going to develop the general theory of k th-order *homogeneous* linear difference equation (3.2.7). We start our exposition by introducing three important definitions.

Definition 3.2.4. The functions $f_1(n)$, $f_2(n)$, $\cdots$, $f_k(n)$ are said to be linearly dependent[2] for $n\geqslant n_0$ if there are k constants a_1, a_2, $\cdots$, a_k, not all zero, such that

$$a_1f_1(n)+a_2f_2(n)+\cdots+a_kf_k(n)=0, n\geqslant n_0 \tag{3.2.12}$$

The negation of linear dependence is linear independence[3].

The functions $f_1(n)$, $f_2(n)$, $\cdots$, $f_k(n)$ are said to be linearly independent

---

[1] initial value problem 初值问题

[2] linearly dependent 线性相关

[3] linear independence 线性无关

for $n \geqslant n_0$, if whenever

$$a_1 f_1(n)+a_2 f_2(n)+\cdots+a_k f_k(n)=0, \text{for all} \quad n \geqslant n_0$$

then we must have $a_1=a_2=\cdots=a_k=0$.

Thus, two functions $f_1(n)$, $f_2(n)$ are linearly dependent if one is a multiple of the other, i. e., $f_1(n)=a f_2(n)$, for some constant a.

【Example 3. 2. 5】 Show that the functions 2^n, $n 2^n$, and $n^2 2^n$ are linearly independent for $n \geqslant 1$.

Solution. Suppose that there are three constants a_1, a_2, a_3, such that

$$a_1 2^n+a_2 n 2^n+a_3 n^2 2^n=0, \text{for all} \quad n \geqslant 1$$

Then by dividing by 2^n, we get

$$a_1+a_2 n+a_3 n^2=0, \text{for all} \quad n \geqslant 1$$

This is impossible unless $a_3=0$, since a second-degree equation in n possesses at most two solutions $n \geqslant 1$. Similarly, $a_2=0$, whence $a_1=0$. Hence $a_1=a_2=a_3=0$, which establishes the linear independence of our functions.

Definition 3. 2. 5. A set of k linearly independent solutions of equation (3. 2. 7) is called a fundamental set of solutions[1].

It is not practical to cheek the linear independence of a set of solutions using the definition. There is a simple method to check the linear independence of solutions using *Casoratian*[2] $W(n)$, which is the discrete analogue of Wronskian in differential equations[3].

Definition 3. 2. 6. The *Casoratian* $W(n)$ of solutions $y_1(n), y_2(n), \cdots, y_k(n)$ is given by

$$W(n)=\det\begin{bmatrix} y_1(n) & y_2(n) & \cdots & y_k(n) \\ E y_1(n) & E y_2(n) & \cdots & E y_k(n) \\ \vdots & \vdots & & \vdots \\ E^{k-1} y_1(n) & E^{k-1} y_2(n) & \cdots & E^{k-1} y_k(n) \end{bmatrix}$$

$$=\det\begin{bmatrix} y_1(n) & y_2(n) & \cdots & y_k(n) \\ y_1(n+1) & y_2(n+1) & \cdots & y_k(n+1) \\ \vdots & \vdots & & \vdots \\ y_1(n+k-1) & y_2(n+k-1) & \cdots & y_k(n+k-1) \end{bmatrix} \tag{3. 2. 13}$$

Next we give a formula to compute the Casoratian $W(n)$.

---

❶ fundamental set of solutions 基本解组

❷ Casoration 卡索拉蒂行列式

❸ discrete analogue of Wronskian 朗斯基（Wronsky，朗斯基，波兰数学家）行列式的离散模拟

Lemma 3. 2. 3. (Abel's Lemma). Let $y_1(n)$, $y_2(n)$, $\cdots$, $y_k(n)$ be the solutions of (3. 2. 7) and $W(n)$ be their *Casoratian*. Then, for $n \geqslant n_0$,

$$W(n)=(-1)^{k(n-n_0)}\left[\prod_{i=n_0}^{n-1} p_k(i)\right] W(n_0) \tag{3. 2. 14}$$

Proof. We will only proof the case $k=3$, since the general case may be established in a similar fashion. So let $y_1(n)$, $y_2(n)$ and $y_3(n)$ be three linearly independent solutions of (3. 2. 7). Then we have

$$W(n+1)=\det\begin{bmatrix} y_1(n+1) & y_2(n+1) & y_3(n+1) \\ y_1(n+2) & y_2(n+2) & y_3(n+2) \\ y_1(n+3) & y_2(n+3) & y_3(n+3) \end{bmatrix} \tag{3. 2. 15}$$

From (3. 2. 7) we have,

$$\begin{aligned} y_1(n+3) &= -p_1(n) y_1(n+2) - p_2(n) y_1(n+1) - p_3(n) y_1(n) \\ y_2(n+3) &= -p_1(n) y_2(n+2) - p_2(n) y_2(n+1) - p_3(n) y_2(n) \\ y_3(n+3) &= -p_1(n) y_3(n+2) - p_2(n) y_3(n+1) - p_3(n) y_3(n) \end{aligned} \tag{3. 2. 16}$$

We use formula (3. 2. 16) to substitute for $y_1(n+3)$, $y_2(n+3)$, $y_3(n+3)$ in the last row of formula (3. 2. 15) , we obtain

$$\begin{aligned} W(n+1) &= \det\begin{bmatrix} y_1(n+1) & y_2(n+1) & y_3(n+1) \\ y_1(n+2) & y_2(n+2) & y_3(n+2) \\ y_1(n+3) & y_2(n+3) & y_3(n+3) \end{bmatrix} \\ &= \det\begin{bmatrix} y_1(n+1) & y_2(n+1) & y_3(n+1) \\ y_1(n+2) & y_2(n+2) & y_3(n+2) \\ -p_3(n) y_1(n) & -p_3(n) y_2(n) & -p_3(n) y_3(n) \end{bmatrix} \\ &= -p_3(n)\det\begin{bmatrix} y_1(n+1) & y_2(n+1) & y_3(n+1) \\ y_1(n+2) & y_2(n+2) & y_3(n+2) \\ y_1(n) & y_2(n) & y_3(n) \end{bmatrix} \\ &= -p_3(n)(-1)^2\det\begin{bmatrix} y_1(n) & y_2(n) & y_3(n) \\ y_1(n+1) & y_2(n+1) & y_3(n+1) \\ y_1(n+2) & y_2(n+2) & y_3(n+2) \end{bmatrix} \end{aligned}$$

Thus

$$W(n+1)=(-1)^3 p_3(n) W(n) \tag{3. 2. 17}$$

which is a linear first-order difference equation .

Using formula (3. 1. 10), the solution of (3. 2. 17) is given by

$$W(n)=\left[\prod_{i=n_0}^{n}(-1)^3 p_3(i)\right] W(n_0)=(-1)^{3(n-n_0)}\left[\prod_{i=n_0}^{n} p_3(i)\right] W(n_0)$$

Corollary 3. 2. 1. Suppose that $p_k(n)\neq 0$ for all $n\geqslant n_0$. Then the Casoratian $W(n)\neq 0$ for all $n\geqslant n_0$ *if and only if* $W(n_0)\neq 0$.

This important corollary follows immediately from formula (3. 2. 14). The main point in the corollary is that either $W(n)=0$ for all $n\geqslant n_0$ or $W(n)\neq 0$ for all $n\geqslant n_0$ (for some $n_0\in Z^+$).

Theorem 3. 2. 2. Let $y_1(n)$, $y_2(n)$, $\cdots$, $y_k(n)$ be a set of solutions of (3. 2. 7). Then $y_1(n)$, $y_2(n)$, $\cdots$, $y_k(n)$ is a fundamental set if and only if for some $n_0\in Z^+$, $W(n_0)\neq 0$.

Proof. Suppose that for some constants a_1, a_2, $\cdots$, a_k and $n_0\in Z^+$,

$$a_1f_1(n)+a_2f_2(n)+\cdots+a_kf_k(n)=0,\quad n\geqslant n_0$$

Then, we can generate the following $k-1$ equations:

$$a_1f_1(n+1)+a_2f_2(n+1)+\cdots+a_kf_k(n+1)=0$$
$$\vdots$$
$$a_1f_1(n+k-1)+a_2f_2(n+k-1)+\cdots+a_kf_k(n+k-1)=0$$

This assemblage may be transcribed as

$$\begin{bmatrix} y_1(n) & y_2(n) & \cdots & y_k(n) \\ y_1(n+1) & y_2(n+1) & \cdots & y_k(n+1) \\ \vdots & \vdots & & \vdots \\ y_1(n+k-1) & y_2(n+k-1) & \cdots & y_k(n+k-1) \end{bmatrix} \begin{bmatrix} a_1 \\ a_2 \\ \vdots \\ a_k \end{bmatrix} = \begin{bmatrix} 0 \\ 0 \\ \vdots \\ 0 \end{bmatrix} \tag{3. 2. 18}$$

Linear algebra tell us equation (3. 2. 18) has only zero solution (i. e., $a_1=a_2=\cdots=a_k=0$) *if and only if* $W(n)\neq 0$ for all $n\geqslant n_0$. By Corollary 3. 2. 1, $W(n)\neq 0$ for all $n\geqslant n_0$ *if and only if* $W(n_0)\neq 0$.

Thus we verify the Theorem 3. 2. 2.

【Example 3. 2. 6】 Show that the functions n, 2^n form a fundamental set solutions of the equation

$$y(n+2)-\frac{3n-2}{n-1}y(n+1)+\frac{2n}{n-1}y(n)=0$$

Solution.

(1) Let us verify that n, 2^n are a solutions of the equation.

By substituting $y(n)=n$ into the equation:

$$n+2-\frac{3n-2}{n-1}(n+1)+\frac{2n}{n-1}n=0$$

Thus, n is a solution of the equation.

By substituting $y(n)=2^n$ into the equation:

$$2^{n+2}-\frac{3n-2}{n-1}2^{n+1}+\frac{2n}{n-1}2^n=2^n\left(2^2-2\,\frac{3n-2}{n-1}+\frac{2n}{n-1}\right)=0$$

Then 2^n is also a solution of the equation.

(2) To affirm the linear independence of these solutions. The Casoratian of n, 2^n is given by

$$W(n)=\det\begin{bmatrix} n & 2^n \\ n+1 & 2^{n+1} \end{bmatrix}$$

Thus

$$W(0)=\det\begin{bmatrix} 0 & 1 \\ 1 & 2 \end{bmatrix}=-1\neq 0$$

Hence by Theorem 3.2.2, the functions n, 2^n are linearly independent and thus form a fundamental set.

Theorem 3.2.3. (The Fundamental Theorem). If $p_k(n)\neq 0$ for all $n\geqslant n_0$, then the equation (3.2.6) has a fundamental set of solutions for $n\geqslant n_0$.

Proof. By Theorem (3.2.1), there are solutions $y_1(n)$, $y_2(n)$, $\cdots$, $y_k(n)$ such that $y_i(n_0+i-1)=1$, $y_i(n_0)=y_i(n_0+1)=\cdots=y_i(n_0+i-2)=y_i(n_0+i)=\cdots=y_i(n_0+k-1)=0$, $1\leqslant i\leqslant k$. Hence $y_1(n_0)=y_2(n_0+1)=\cdots=y_k(n_0+k-1)=1$. It follows that $W(n_0)=\det I=1$. By Theorem 3.2.2 the set $\{y_1(n), y_2(n), \cdots, y_k(n)\}$ is a fundamental set of solutions of equation (3.2.6).

There are infinitely many fundamental sets of solutions of (3.2.6). The next Lemma presents a method of generating fundamental sets starting from a known set.

Lemma 3.2.4. Let $y_1(n)$, $y_2(n)$ be two solutions of (3.2.6). Then the following statements hold:

(1) $y(n)=y_1(n)+y_2(n)$ is a solution of equation (3.2.6).

(2) $\tilde{y}(n)=ay_1(n)$ is a solution of equation (3.2.6) for any constant a.

(Exercises 2.7, Problem 10).

From Lemma 3.2.4, we conclude the following principle.

Superposition Principle[1]. If $y_1(n)$, $y_2(n)$, $\cdots$, $y_k(n)$ are solutions of equation (3.2.6), then

$$y(n)=a_1y_1(n)+a_2y_2(n)+\cdots+a_ky_k(n)$$

is also a solution of equation (3.2.6).

(Exercises 2.7, Problem 11).

Now let $\{y_1(n), y_2(n), \cdots, y_k(n)\}$ be a fundamental set of solutions of equation (3.2.6) and $y(n)$ be any given solution of equation (3.2.6). Then there are constants

[1] Superposition Principle 叠加原理

a_1, a_2, $\cdots$, a_k, such that $y(n)=\sum_{i=1}^{k} a_i y_i(n)$. To show this, we write

$$\begin{bmatrix} y_1(n) & y_2(n) & \cdots & y_k(n) \\ y_1(n+1) & y_2(n+1) & \cdots & y_k(n+1) \\ \vdots & \vdots & & \vdots \\ y_1(n+k-1) & y_2(n+k-1) & \cdots & y_k(n+k-1) \end{bmatrix} \begin{bmatrix} a_1 \\ a_2 \\ \vdots \\ a_k \end{bmatrix} = \begin{bmatrix} x(n) \\ x(n+1) \\ \vdots \\ x(n+k-1) \end{bmatrix}$$

Since $W(n) \neq 0$, it follows that

$$\begin{bmatrix} a_1 \\ a_2 \\ \vdots \\ a_k \end{bmatrix} = \begin{bmatrix} y_1(n) & y_2(n) & \cdots & y_k(n) \\ y_1(n+1) & y_2(n+1) & \cdots & y_k(n+1) \\ \vdots & \vdots & & \vdots \\ y_1(n+k-1) & y_2(n+k-1) & \cdots & y_k(n+k-1) \end{bmatrix}^{-1} \begin{bmatrix} x(n) \\ x(n+1) \\ \vdots \\ x(n+k-1) \end{bmatrix}$$

The above discussion leads us to define the general solution of (3.2.6).

Definition 3.2.7. Let the set $\{y_1(n), y_2(n), \cdots, y_k(n)\}$ be a fundamental set of solutions of equation (3.2.6). Then the general solution[1] of equation (3.2.6) is given by $y(n)=\sum_{i=1}^{k} a_i y_i(n)$, for arbitrary constants a_i.

3.3 Linear Homogeneous equations with constant coefficients

In this section, we focus on solving the kth-order difference equation

$$y(n+k)+p_1 y(n+k-1)+\cdots+p_k y(n)=0 \tag{3.3.1}$$

where the p_i's are constants and $p_k \neq 0$. Our objective is to find a fundamental set of solutions and the general solution of equation (3.3.1).

We suppose that solutions of equation (3.3.1) are in the form λ^n, where λ is a complex number. Substituting this value into equation (3.3.1), we obtain

$$\lambda^k + p_1 \lambda^{k-1} + \cdots + p_{k-1}\lambda + p_k = 0 \tag{3.3.2}$$

(3.3.2) is called the characteristic equation[2] of equation (3.3.1), and its roots are called the characteristic roots[3]. Notice that since $p_k \neq 0$, it is easy to prove that $\lambda \neq 0$.

Case (a) The *characteristic roots* λ_1, λ_2, $\cdots$, λ_k are distinct.

It is easy to show that the set $\{\lambda_1{}^n, \lambda_2{}^n, \cdots, \lambda_k{}^n\}$ is a set of solutions. We are going to prove that the set $\{\lambda_1{}^n, \lambda_2{}^n, \cdots, \lambda_k{}^n\}$ is a fundamental set of solu-

---

[1] general solution 通解

[2] characteristic equation 特征方程

[3] characteristic roots 特征根

tions. To prove this, it is suffices to show that $W(0) \neq 0$, where $W(n)$ is the Casoration of the solutions. That is

$$W(0)=\det\begin{bmatrix} 1 & 1 & \cdots & 1 \\ \lambda_1^1 & \lambda_2^1 & \cdots & \lambda_k^1 \\ \lambda_1^2 & \lambda_2^2 & \cdots & \lambda_k^2 \\ \vdots & \vdots & & \vdots \\ \lambda_1^{k-1} & \lambda_2^{k-1} & \cdots & \lambda_k^{k-1} \end{bmatrix} \tag{3.3.3}$$

The determinant (3.3.3) is called the Vandermonde determinant[1]. It may be shown by mathematical induction that

$$W(0)=\prod_{1 \leqslant i < j \leqslant k}(\lambda_j-\lambda_i) \tag{3.3.4}$$

Since the *characteristic roots* $\lambda_1, \lambda_2, \cdots, \lambda_k$ are distinct, it follows that $W(0) \neq 0$. By Theorem 3.2.2, the set $\{\lambda_1{}^n, \lambda_2{}^n, \cdots, \lambda_k{}^n\}$ is a fundamental set of solutions. Consequently, the general solution of equation (3.3.1) is

$$y(n)=\sum_{i=1}^{k} a_i \lambda_i{}^n \tag{3.3.5}$$

where a_i's are complex numbers.

Case (b) The *characteristic roots* $\lambda_1, \lambda_2, \cdots, \lambda_r$ are distinct with multiplicities $m_1, m_2, \cdots, m_r$, where $m_1+m_2+\cdots+m_r=k$.

In this case (3.3.1) may be written as

$$(E-\lambda_1)^{m_1}(E-\lambda_2)^{m_2}\cdots(E-\lambda_r)^{m_r}y(n)=0 \tag{3.3.6}$$

If $\varphi(n)$ is a solution of

$$(E-\lambda_i)^{m_i}y(n)=0 \tag{3.3.7}$$

then it is also a solution of equation (3.3.6).

Lemma 3.3.1. If λ_i is a characteristic root with multiplicity m_i, then the set $G_i=\{\lambda_i^n, C_n^1\lambda_i^{n-1}, C_n^2\lambda_i^{n-2}, \cdots, C_n^{m_i-1}\lambda_i^{n-m_i+1}\}$ is a fundamental set of solutions of (3.3.7), where $C_n^r=\dfrac{n(n-1)\cdots(n-r+1)}{r!}$.

Proof. We need to show that $C_n^r\lambda_i^{n-r}$ is a solution of (3.3.7) for all $0 \leqslant r \leqslant m_i-1$. From the Lemma 3.2.2, it follows that

$$\begin{aligned}(E-\lambda_i)^{m_i}C_n^r\lambda_i^{n-r} &= \lambda_i^{n-r}(\lambda_i E-\lambda_i)^{m_i}C_n^r \\ &= \lambda_i^{n+m_i-r}(E-I)^{m_i}C_n^r \\ &= \lambda_i^{n+m_i-r}\Delta^{m_i}C_n^r\end{aligned}$$

Since the degree of polynomial C_n^r is r, it follows from equation (3.3.5) that

---

[1] Vandermonde determinant 范德蒙德行列式

$$\Delta^{m_i} C_n^r = 0, \ (E-\lambda_i)^{m_i} C_n^r \lambda_i^{n-r} = 0$$

To show that G_i is a fundamental set of solutions of (3. 3. 7), it needs to show that $W(n) \neq 0$.

As

$$W(0) = \det \begin{pmatrix} 1 & 0 & \cdots & 0 \\ \lambda_i & 1 & \cdots & 0 \\ \lambda_i^2 & 2\lambda_i & \cdots & 0 \\ \vdots & \vdots & & \vdots \\ \lambda_i^{m_i-1} & \frac{m_i-1}{1!}\lambda_i^{m_i-2} & \cdots & 1 \end{pmatrix} = 1 \neq 0$$

by Corollary 3. 2. 1, we have $W(n) \neq 0$.

Theorem 3. 3. 1. The set $G = \bigcup_{i=1}^{r} G_i$ is a fundamental set of solutions of equation (3. 3. 1).

Proof. By Lemma 3. 3. 1, the functions in G are solutions of equation (3. 3. 1). Now

$$W(0) = \det \begin{bmatrix} 1 & 0 & \cdots & 1 & 0 & \cdots \\ \lambda_1 & 1 & \cdots & \lambda_r & 1 & \cdots \\ \lambda_1^2 & 2\lambda_1 & \cdots & \lambda_r^2 & 2\lambda_r & \cdots \\ \vdots & \vdots & & \vdots & \vdots & \\ \lambda_1^{k-1} & (k-1)\lambda_1^{k-2} & \cdots & \lambda_r^{k-1} & (k-1)\lambda_r^{k-2} & \cdots \end{bmatrix} \qquad (3.\ 3.\ 8)$$

The determinant is called the generalized Vandermonde determinant. It may be shown that

$$W(0) = \prod_{1 \leqslant i < j \leqslant k} (\lambda_j - \lambda_i)^{m_j m_i} \qquad (3.\ 3.\ 9)$$

The *characteristic roots* λ_1, λ_2, $\cdots$, λ_r are distinct, then $W(0) \neq 0$. Hence by Corollary 3. 2. 1, we have $W(n) \neq 0$. Thus by Theorem 3. 2. 2, G is a fundamental set of solutions of equation (3. 3. 1).

From Lemma 3. 3. 1 and Theorem 3. 3. 1, we conclude the following corollary.

Corollary 3. 3. 1. The general solution of equation (3. 3. 1) is given by

$$y(n) = \sum_{i=1}^{r} \lambda_i^n (a_{i0} + a_{i1} n + \cdots + a_{i,m_i-1} n^{m_i-1}) \qquad (3.\ 3.\ 10)$$

【Example 3. 3. 1】 Solve the equation

$$y(n+3) - 7y(n+2) + 16y(n+1) - 12y(n) = 0$$

$$y(0) = 0, \quad y(1) = 1, \quad y(2) = 1$$

Solution. The characteristic equation is

$$\lambda^3 - 7\lambda^2 + 16\lambda - 12 = 0$$

Thus, the characteristic roots are $\lambda_1 = \lambda_2 = 2$, $\lambda_3 = 3$.

From the formula (3.3.10), the general solution is

$$y(n)=a_0 2^n+a_1 n 2^n+b_1 3^n$$

Using the initial data, we obtain the following system of equations

$$y(0)=a_0+b_1=0$$

$$y(1)=2a_0+2a_1+3b_1=1$$

$$y(2)=4a_0+8a_1+9b_1=1$$

After solving the system of equations, we can find that

$$a_0=3,\quad a_1=2,\quad b_1=-3$$

Hence the solution of the equation is given by $y(n)=3\cdot 2^n+2n2^n-3\cdot 3^n$.

Case (c) Complex Characteristic roots

In this case, we only consider the second-order equation $y(n+2)+p_1 y(n+1)+p_2 y(n)=0$. Suppose that $\lambda_1=\alpha+\mathrm{i}\beta$, $\lambda_2=\alpha-\mathrm{i}\beta$ are the complex roots of the characteristic equation, then the general solution would be

$$y(n)=c_1(\alpha+\mathrm{i}\beta)^n+c_2(\alpha-\mathrm{i}\beta)^n$$

In polar coordinates,

$$\alpha=r\cos\theta,\quad \beta=r\sin\theta,\quad r=\sqrt{\alpha^2+\beta^2},\quad \theta=\arctan\left(\frac{\beta}{\alpha}\right)$$

By using De Moivre's Theorem[1],

$$[r(\cos\theta+\mathrm{i}\sin\theta)]^n=r^n(\cos n\theta+\mathrm{i}\sin n\theta)$$

it follows that

$$\begin{aligned}
y(n)&=c_1(\alpha+\mathrm{i}\beta)^n+c_2(\alpha-\mathrm{i}\beta)^n\\
&=c_1(r\cos\theta+\mathrm{i}r\sin\theta)^n+c_2(r\cos\theta-\mathrm{i}r\sin\theta)^n\\
&=c_1r^n(\cos n\theta+\mathrm{i}\sin n\theta)+c_2r^n(\cos n\theta-\mathrm{i}\sin n\theta)\\
&=r^n[(c_1+c_2)\cos n\theta+\mathrm{i}(c_1-c_2)\sin n\theta]\\
&=r^n(a_1\cos n\theta+a_2\sin n\theta)
\end{aligned}\tag{3.3.11}$$

where $a_1=c_1+c_2$, and $a_2=\mathrm{i}(c_1-c_2)$.

Let

$$\cos\omega=\frac{a_1}{\sqrt{a_1^2+a_2^2}},\ \sin\omega=\frac{a_2}{\sqrt{a_1^2+a_2^2}},\ \omega=\arctan\left(\frac{a_2}{a_1}\right)$$

Then formula (3.3.11) becomes

$$\begin{aligned}
y(n)&=r^n\sqrt{a_1^2+a_2^2}[\cos\omega\cos n\theta+\sin\omega\sin n\theta]\\
&=r^n\sqrt{a_1^2+a_2^2}\cos(n\theta-\omega)\\
&=Ar^n\cos(n\theta-\omega)
\end{aligned}$$

---

[1] 棣莫弗(Abraham De Moivre),法国裔英国籍数学家

3.4 Linear Nonhomogeneous equations

In this section we focus on solving the kth-order linear nonhomogeneous equation

$$y(n+k)+p_1(n)y(n+k-1)+\cdots+p_k(n)y(n)=g(n) \tag{3.4.1}$$

where $p_k(n)\neq 0$ for all $n\geqslant 0$. The sequence $g(n)$ is called the forcing term, the control, or the input of the system.

【Example 3.4.1】 Consider the equation

$$y(n+2)-y(n-1)-6y(n)=5\cdot 3^n$$

(1) Show that $y_1(n)=n3^{n-1}$ and $y_2(n)=(1+n)\ 3^{n-1}$ are solutions of the equation.

(2) Show that $y(n)=y_2(n)-y_1(n)$ is not a solution of the equation.

(3) Show that $\varphi(n)=cy_1(n)$ is not a solution of the equation, where $c\neq 1$ is a constant.

Solution.

(1) Substituting $y_1(n)=n3^{n-1}$ and $y_2(n)=(n+1)\ 3^{n-1}$ into the equation gives

$$(n+2)3^{n+1}-(n+1)3^n-6n3^{n-1}=3^{n-1}(9n+18-3n-3-6n)=5\cdot 3^n$$

$$(n+3)3^{n+1}-(n+2)3^n-6(n+1)3^{n-1}=3^{n-1}(9n+27-3n-6-6n-6)=5\cdot 3^n$$

Hence $y_1(n)=n3^{n-1}$ and $y_2(n)=(n+1)3^{n-1}$ are solutions of the equation.

(2) $y(n)=y_2(n)-y_1(n)=3^{n-1}$. Substituting this into the equation gives

$$3^{n+1}-3^n-6\cdot 3^{n-1}=3^n(3-1-2)=0\neq 5\cdot 3^n$$

Thus, $y(n)=y_2(n)-y_1(n)$ is not a solution of the equation.

(3) By substituting $\varphi(n)$ into the equation we know that $\varphi(n)$ is not a solution of the equation.

From the above example we conclude that neither the sum (difference) of two solutions nor a multiple of a solution is a solution. In general, solutions of the nonhomogeneous equation (3.4.1) do not form a vector space.

Theorem 3.4.1. If $y_1(n)$ and $y_2(n)$ are solutions of the equation (3.4.1), the $y_2(n)-y_1(n)$ is a solution of the corresponding homogeneous equation

$$y(n+k)+p_1(n)y(n+k-1)+\cdots+p_k(n)y(n)=0 \tag{3.4.2}$$

Theorem 3.4.2. Any solution $y(n)$ of equation (3.4.1) may be written as

$$y(n)=y_p(n)+y_c(n)=y_p(n)+\sum_{i=1}^{k}a_iy_i(n) \tag{3.4.3}$$

Where $y_p(n)$ is a particular solution[1] of the nonhomogeneous equation (3.4.1),

---

[1] particular solution 特解

$\{y_1(n), y_2(n), \cdots, y_k(n)\}$ is a fundamental set of solutions of the homogeneous equation (3.4.2), and $y_c(n) = \sum_{i=1}^{k} a_i y_i(n)$ is the general solution of equation (3.4.2).

Proof. According to Theorem 3.4.1, $y(n) - y_p(n)$ is a solution of the homogeneous equation (3.4.2). Thus $y(n) - y_p(n) = \sum_{i=1}^{k} a_i y_i(n)$, for some constants a_i.

Now we focus our attention on finding a particular solution $y_p(n)$ of the nonhomogeneous equation with constant coefficients such as

$$y(n+k) + p_1 y(n+k-1) + \cdots + p_k y(n) = g(n) \tag{3.4.4}$$

We use the method of *undetermined coefficients*[1] to compute $y_p(n)$. The method consists in making an intelligent guess as to the form of the particular solution and then substituting this function into the difference equation. But for a completely arbitrary nonhomogeneous term $g(n)$, this method is not effective. However, definite rules can be established for the determination of a particular solution by this method when $g(n)$ is a linear combination of terms, each having one of the forms

$$a^n, \sin(bn), \cos(bn), \text{or} \quad n^k \tag{3.4.5}$$

or products of these forms such as

$$a^n \sin(bn), a^n \cos(bn), a^n n^k, a^n n^k \sin(bn), \cdots \tag{3.4.6}$$

Definition 3.4.1. A polynomial operator $N(E)$ is said to be an annihilator[2] of $g(n)$ if

$$N(E) g(n) = 0 \tag{3.4.7}$$

where E is the shift operator.

For example, $N(E) = E - 3$ is an annihilator of $g(n) = 3^n$, since $(E-3)3^n = 0$.

By using shift operator E, the equation (3.4.4) becomes

$$P(E) y(n) = g(n) \tag{3.4.8}$$

Where $P(E) = E^k + p_1 E^{k-1} + p_2 E^{k-2} + \cdots + p_k I$.

Assume that $N(E)$ is an annihilator of $g(n)$ in (3.4.8). Applying $N(E)$ on both sides of (3.4.8), we obtain

$$N(E) P(E) y(n) = N(E) g(n) = 0 \tag{3.4.9}$$

Let $\lambda_1, \lambda_2, \cdots, \lambda_k$ be the characteristic roots of

$$P(E) y(n) = 0 \tag{3.4.10}$$

and $\mu_1, \mu_2, \cdots, \mu_k$ be the characteristic roots of

---

[1] undetermined coefficients 待定系数

[2] annihilator 零花算子

$$N(E)y(n)=0 \tag{3.4.11}$$

We must consider two separate cases.

Case (a). None of the λ_i's equals any of the μ_i's. In this case, write $y_p(n)$ as the general solution of (3.4.11) with undetermined constants. Substituting this particular solution into (3.4.4), we find the values of the constants. Table 3.1 contains several types of functions $g(n)$ and their corresponding particular solutions.

Case (b). If $\lambda_i=\mu_j$ for some i, j. In this case, the set of characteristic roots of equation (3.4.9) is equal to the union of the sets $\{\lambda_1, \lambda_2, \cdots, \lambda_k\}$ $\{\mu_1, \mu_2, \cdots, \mu_k\}$. Consequently, it contains roots of higher multiplicity than the two individual sets of characteristic roots. To determine a particular solution $y_p(n)$, we first find the general solution of equation (3.4.9) and then drop all the terms that appear in $y_c(n)$. Then proceed as in the Case (a) to evaluate the constants.

Table 3.4.1 Particular solutions $y_p(n)$

$g(n)$	$y_p(n)$
a^n	c_1a^n
n^k	$c_0+c_1n+\cdots+c_kn^k$
a^nn^k	$a^n(c_0+c_1n+\cdots+c_kn^k)$
$\sin(bn)$, $\cos(bn)$	$c_1\sin(bn)+c_2\cos(bn)$
$a^n\sin(bn)$, $a^n\cos(bn)$	$a^n[c_1\sin(bn)+c_2\cos(bn)]$
$a^nn^k\sin(bn)$, $a^nn^k\cos(bn)$	$(c_0+c_1n+\cdots+c_kn^k)a^n\sin(bn)$ $+(d_0+d_1n+\cdots+d_kn^k)a^n\cos(bn)$

【Example 3.4.2】 Solve the difference equation

$$y(n+2)+y(n+1)-12y(n)=n2^n$$

Solution. The characteristic roots of the homogeneous equation are $\lambda_1=3$, $\lambda_2=-4$. Hence,

$$y_c(n)=c_13^n+c_2(-4)^n$$

Since the annihilator of $n2^n$ is $N(E)=(E-2)^2$, we know that $\mu_1=\mu_2=2$. Since $\lambda_i\neq\mu_j$, for any i, j. So we let

$$y_p(n)=a_12^n+a_2n2^n$$

Substituting this relation into the equation gives

$$a_12^{n+2}+a_2(n+2)2^{n+2}+a_12^{n+1}+a_2(n+1)2^{n+1}-12a_12^n-12a_2n2^n=n2^n$$

It can be written as

$$(10a_2-6a_1)2^n-6a_2n2^n=n2^n$$

Hence

$$10a_2-6a_1=0,\ -6a_2=1$$

or

$$a_1=-\frac{5}{18},\ a_2=-\frac{1}{6}$$

Thus

$$y_p(n)=-\frac{5}{18}2^n-\frac{1}{6}n2^n$$

and the general solution is

$$y(n)=c_1 3^n+c_2(-4)^n-\frac{5}{18}2^n-\frac{1}{6}n2^n$$

【Example 3. 4. 3】 Solve the difference equation

$$y(n+2)-y(n+1)-6y(n)=5\cdot 3^n$$

Solution. The characteristic roots of the homogeneous equation are $\lambda_1=3$, $\lambda_2=-2$. Hence,

$$y_c(n)=c_1 3^n+c_2(-2)^n$$

The annihilator of $5(3^n)$ is $N(E)=(E-3)$, we know that $\mu_1=3$. Since $\lambda_1=\mu_1$, so we let

$$y_p(n)=c_1 n 3^n$$

Substituting this relation into the equation gives

$$c_1(n+2)3^{n+2}-c_1(n+1)3^{n+1}+6c_1 n 3^n=5\cdot 3^n$$

or

$$c_1=\frac{1}{3}$$

Thus

$$y_p(n)=\frac{1}{3}n3^n=n3^{n-1}$$

And the general solution of the equation is

$$y(n)=c_1 3^n+c_2(-2)^n+n3^{n-1}$$

【Example 3. 4. 4】 Solve the difference equation

$$y(n+2)+4y(n)=8\cdot 2^n\cdot\cos\left(\frac{n\pi}{2}\right)$$

Solution. The characteristic equation of homogeneous equation is

$$\lambda^2+4=0$$

The characteristic roots are

$$\lambda_1=2\mathrm{i},\ \lambda_2=-2\mathrm{i}$$

Thus $r=2$, $\theta=\frac{\pi}{2}$, and

$$y_c(n)=2^n\left(c_1\cos\left(\frac{n\pi}{2}\right)+c_2\sin\left(\frac{n\pi}{2}\right)\right)$$

Since $g(n)=8\cdot 2^n\cdot\cos\left(\frac{n\pi}{2}\right)$, we set

$$y_p(n)=2^n\left(cn\cos\left(\frac{n\pi}{2}\right)+dn\sin\left(\frac{n\pi}{2}\right)\right)$$

Substituting this $y_p(n)$ relation into the equation gives

$$2^{n+2}\left[c(n+2)\cos\left(\frac{n\pi}{2}+\pi\right)+d(n+2)\sin\left(\frac{n\pi}{2}+\pi\right)\right]+4\cdot 2^n\left[cn\cos\left(\frac{n\pi}{2}\right)+dn\sin\left(\frac{n\pi}{2}\right)\right]$$

$$=8\cdot 2^n\cdot\cos\left(\frac{n\pi}{2}\right)-8c\cdot 2^n\cos\left(\frac{n\pi}{2}\right)+4d\cdot 2^n(2n+2)\sin\left(\frac{n\pi}{2}\right)$$

$$=8\cdot 2^n\cdot\cos\left(\frac{n\pi}{2}\right)$$

Hence $c=-1$, $d=0$.

Thus

$$y_p(n)=-2^n n\cos(\frac{n\pi}{2})$$

And the general solution of the equation is

$$y(n)=2^n\left(c_1\cos\left(\frac{n\pi}{2}\right)+c_2\sin\left(\frac{n\pi}{2}\right)-n\cos\left(\frac{n\pi}{2}\right)\right)$$

3.5 Limiting behavior of solution[1]

To simplify our exposition we restrict our discussion to the second-order difference equation

$$y(n+2)+p_1y(n+1)+p_2y(n)=0 \tag{3.5.1}$$

Suppose that λ_1, λ_2 are the characteristic roots of the equation. Then we have three cases.

Case (a): λ_1, λ_2 are distinct real roots.

Then $y_1(n)=\lambda_1{}^n$, $y_2(n)=\lambda_2^n$ are two linearly independent solutions of (3.5.1). We will show that the limiting behavior of the general solution $y(n)=c_1\lambda_1{}^n+c_2\lambda_2^n$ is determined by the behavior of the dominant solution. Without loss of generality we assume that $|\lambda_1|>|\lambda_2|$. Then

$$y(n)=\lambda_1{}^n\left[c_1+c_2\left(\frac{\lambda_2}{\lambda_1}\right)^n\right]$$

---

[1] limiting behavior of solution 解的极限性态

Since

$$\left(\frac{\lambda_1}{\lambda_2}\right)^n \to 0 \quad \text{as} \quad n \to \infty$$

it follows that $\lim\limits_{n\to\infty} y(n) = \lim\limits_{n\to\infty} \lambda_1{}^n \left[c_1 + c_2\left(\frac{\lambda_2}{\lambda_1}\right)^n\right] = \lim\limits_{n\to\infty} c_1\lambda_1{}^n$. There are six different situations that may arise here depending on the value of λ_1.

(1) $\lambda_1 > 1$: The sequence $\{c_1\lambda_1{}^n\}$ diverges and

$$\lim_{n\to\infty} y(n) = \lim_{n\to\infty} c_1\lambda_1{}^n = \infty \text{(unstable system)}$$

(2) $\lambda_1 = 1$: The sequence $\{c_1\lambda_1{}^n\}$ is a constant sequence and

$$\lim_{n\to\infty} y(n) = \lim_{n\to\infty} c_1\lambda_1{}^n = c_1$$

(3) $0 < \lambda_1 < 1$: The sequence $\{c_1\lambda_1{}^n\}$ is monotonically decreasing to zero and

$$\lim_{n\to\infty} y(n) = \lim_{n\to\infty} c_1\lambda_1{}^n = 0 \text{(stable system)}$$

(4) $-1 < \lambda_1 < 0$: The sequence $\{c_1\lambda_1{}^n\}$ is oscillating around zero and converging to zero (stable system).

(5) $\lambda_1 = -1$: The sequence $\{c_1\lambda_1{}^n\}$ is oscillating between two values c_1 and c_2.

(6) $\lambda_1 < -1$: The sequence $\{c_1\lambda_1{}^n\}$ is oscillating but increasing in magnitude (unstable system).

Case (b): $\lambda_1 = \lambda_2 = \lambda$.

The general solution of equation (3.5.1) is $y(n) = (c_1 + c_2 n)\lambda^n$.

(1) $|\lambda| \geq 1$: The sequence $y(n)$ diverges.

(2) $|\lambda| < 1$: $\lim\limits_{n\to\infty} y(n) = \lim\limits_{n\to\infty} (c_1 + c_2 n)\lambda^n = 0$.

Case (c): $\lambda_1 = \alpha + i\beta$ and $\lambda_2 = \alpha - i\beta$, where $\beta \neq 0$.

The general solution of equation (3.5.1) is $y(n) = ar^n \cos(n\theta - \omega)$, where

$$r = \sqrt{\alpha^2 + \beta^2},\ \theta = \arctan\left(\frac{\beta}{\alpha}\right)$$

Since the cosine function oscillates, so that the solution $y(n) = ar^n \cos(n\theta - \omega)$ oscillates.

(1) $r > 1$: Here λ_1 and λ_2 are outside the unit circle. Hence the sequence $y(n)$ is oscillating but increasing in magnitude (unstable system).

(2) $r = 1$: Here λ_1 and λ_2 are lie on the unit circle. Hence the sequence $y(n)$ is oscillating but constant in magnitude.

(3) $r < 1$: Here λ_1 and λ_2 are inside the unit circle. Hence the sequence $y(n)$ is oscillating but converges to zero as $n \to \infty$ (stable system).

Finally, we summarize the above discussion in the following theorem.

Theorem 3.5.1 The following statements hold:

(1) All solution of equation (3.5.1) oscillate if and only if the characteristic equation has no positive real roots.
(2) All solution of equation (3.5.1) converge to zero if and only if $\max\{|\lambda_1|, |\lambda_2|\}<1$.
Next we consider nonhomogeneous difference equations

$$y(n+2)+p_1y(n+1)+p_2y(n)=M \tag{3.5.2}$$

where M is a nonzero constant. we have the equilibrium point or solution $y(n)=y^*$. From equation (3.5.2) we have $y^*+p_1y^*+p_2y^*=M$, or $y^*=\frac{M}{1+p_1+p_2}$ is a particular solution of equation (3.5.2). Consequently, the general solution of equation (3.5.2) is given by

$$y(n)=y^*+y_c(n) \tag{3.5.3}$$

Theorem 3.5.2 The following statements hold:
(1) All solution of equation (3.5.2) oscillate about y^* if and only if the characteristic equation of equation (3.5.1) has no positive real roots.
(2) All solution of equation (3.5.2) converge to y^* if and only if $\max\{|\lambda_1|, |\lambda_2|\}<1$, where λ_1 and λ_2 are the characteristic roots of equation (3.5.1).

3.6 Autonomous (Time-Invariant) Systems[1]

In this section we are interested in finding solutions of the following system of linear equations[2]:

$$\begin{cases} y_1(n+1)=a_{11}y_1(n)+a_{12}y_2(n)+\cdots+a_{1k}y_k(n) \\ y_2(n+1)=a_{21}y_1(n)+a_{22}y_2(n)+\cdots+a_{2k}y_k(n) \\ \cdots \\ y_k(n+1)=a_{k1}y_1(n)+a_{k2}y_2(n)+\cdots+a_{kk}y_k(n) \end{cases} \tag{3.6.1}$$

The system may be written in vector form

$$\mathbf{y}(n+1)=\mathbf{A}\mathbf{y}(n)$$

where $\mathbf{y}(n)=(y_1(n), y_2(n), \cdots, y_k(n))^{\mathrm{T}}\in R^k$, and $\mathbf{A}=(a_{ij})_{k\times k}$ is a real matrix.

Then $\mathbf{y}(n)=\mathbf{A}^n\mathbf{y}(0)$ is the solution of (3.6.1).

3.7 Exercises

1. Find the solution of each difference equation:

[1] autonomous (time-invariant) Systems 自治（时不变）系统
[2] system of linear equations 线性差分方程组

(1) $x(n+1)=2x(n)+1$, $x(0)=c$.

(2) $x(n+1)=2x(n)-2$, $x(0)=\frac{1}{2}$.

(3) $x(n+1)-3x(n)=1$, $x(0)=1$.

(4) $x(n+1)+\frac{1}{2}x(n)=3$, $x(0)=1$.

2. Find the solution of each difference equation:

(1) $x(n+1)-(n+1)x(n)=0$, $x(0)=c$.

(2) $x(n+1)-3^n x(n)=0$, $x(0)=c$.

(3) $x(n+1)-e^{2n}x(n)=0$, $x(0)=c$.

(4) $x(n+1)+\frac{n}{n+1}x(n)=0$, $n\geqslant 1$ $x(0)=c$.

3. Show that the operators Δ and E are linear.

4. Show that $E^k x(n)=\sum_{i=0}^{k} C_k^i \Delta^{k-i} x(n)$.

5. Prove the Lemma 3. 2. 1.

6. Prove the Lemma 3. 2. 2.

7. Find the Casoratian of the following functions and determine whether they are linearly dependent or independent:

(1) 2^n, $3\cdot 2^{n+2}$, e^n.

(2) 3^n, $n\cdot 3^n$, $n^2 3^n$.

(3) 2^n, $(-2)^n$, 3.

(4) 0, 3^n, 7^n.

8. Find the Casoratian $W(n)$ of the solutions of the difference equations:

(1) $y(n+3)-10y(n+2)+31y(n+1)-30y(n)=0$, if $W(0)=6$.

(2) $y(n+3)-3y(n+2)+4y(n+1)-12y(n)=0$, if $W(0)=26$.

9. For the following difference equations and their accompanied solutions:

(1) determine whether these solutions are linearly independent

(2) find , if possible, using only the given solutions, the general solution

(a) $y(n+3)-3y(n+2)+3y(n+1)-y(n)=0$; 1, n, n^2,

(b) $y(n+2)+y(n)=0$; $\cos\left(\frac{n\pi}{2}\right)$, $\sin\left(\frac{n\pi}{2}\right)$,

(c) $y(n+3)+y(n+2)-8y(n+1)-12y(n)=0$; 3^n, $(-2)^n$, $(-2)^{n+3}$,

(d) $y(n+4)-16y(n)=0$; 2^n, $n2^n$, $n^2 2^n$.

10. Prove Lemma 3. 2. 4.

11. Prove the superposition principle: If $y_1(n)$, $y_2(n)$, …, $y_k(n)$ are solutions of equation (3. 2. 6), then

$$y(n)=a_1y_1(n)+a_2y_2(n)+\cdots+a_ky_k(n)$$

is also a solution of equation (3.2.6).

12. Find homogeneous difference equations whose solutions are:

(1) $2^{n-1}-5^{n+1}$.

(2) $3\cos\left(\frac{n\pi}{2}\right)-\sin\left(\frac{n\pi}{2}\right)$.

(3) $(n+2)5^n\sin\left(\frac{n\pi}{4}\right)$.

(4) $(c_1+c_2n+c_3n^2)\ 7^n$.

(5) $1+3n-5n^2+6n^3$.

13. Write the general solution of the following equations

(1) $y(n+2)-16y(n)=0$.

(2) $y(n+2)+16y(n)=0$.

(3) $(E-3I)^2\ (E^2+4I)y(n)=0$.

(4) $\Delta^3y(n)=0$.

(5) $(E^2+2I)^2y(n)=0$.

14. Suppose that $y(n+2)+p_1y(n+1)+p_2y(n)=0$ has two complex characteristic roots

$$\lambda_1=\alpha+i\beta, \lambda_2=\alpha-i\beta\ ,\ \text{where}\ \alpha=r\cos\theta, \beta=r\sin\theta.$$

Show that $y_1(n)=r^n\cos n\theta$ and $y_2(n)=r^n\sin n\theta$ are two linearly independent solutions of the given equations.

15. Find particular solutions of the following difference equations.

(1) $y(n+2)-5y(n+1)+6y(n)=1+n$.

(2) $y(n+2)+8y(n+1)+12y(n)=e^n$.

(3) $y(n+2)-5y(n+1)+4y(n)=4^n-n^2$.

(4) $y(n+2)+8y(n+1)+7y(n)=ne^n$.

(5) $y(n+2)-y(n)=n\cos\left(\frac{n\pi}{2}\right)$.

16. Find the solutions of the following difference equations.

(1) $\Delta^2y(n)=16$, $y(0)=2$, $y(1)=3$.

(2) $(\Delta^2+7I)y(n)=2\sin\left(\frac{n\pi}{4}\right)$, $y(0)=0$, $y(1)=1$.

(3) $(E^2+I)(E-3I)y(n)=3^n$, $y(0)=0$, $y(1)=1$, $y(2)=3$.

17. Find the general solutions of the following difference equations.

(1) $y(n+2)-y(n)=n2^n\sin\left(\frac{n\pi}{2}\right)$.

(2) $y(n+2)+8y(n+1)+7y(n)=n2^n$.

18. Determine the stability of the equilibrium points by using Theorem 3. 5. 1 or Theorem 3. 5. 2.

(1) $y(n+2)-2y(n+1)+2y(n)=0$.

(2) $y(n+2)+\frac{1}{4}y(n)=0$.

(3) $y(n+2)-5y(n+1)+6y(n)=0$.

(4) $y(n+2)+y(n+1)+\frac{1}{2}y(n)=-5$.

19. Determine the oscillatory behavior of the solution of the equation.

(1) $y(n+2)-2y(n+1)+y(n)=0$.

(2) $y(n+2)+\frac{1}{4}y(n)=\frac{5}{4}$.

(3) $y(n+2)+5y(n+1)+6y(n)=12$.

(4) $y(n+2)+y(n+1)+\frac{1}{2}y(n)=5$.

20. Solve the syetem

(1) $\begin{cases} y_1(n+1)=-y_1(n)+y_2(n), & y_1(0)=1, \\ y_2(n+1)=2y_2(n), & y_2(0)=2. \end{cases}$

(2) $\begin{cases} y_1(n+1)=y_2(n), \\ y_2(n+1)=y_3(n), \\ y_3(n+1)=2y_1(n)-y_2(n)+y_3(n). \end{cases}$

(3) $y(n+1)=\begin{bmatrix} 1 & -2 & -2 \\ 0 & 0 & -1 \\ 0 & 2 & 3 \end{bmatrix} y(n)$, $y(0)=\begin{bmatrix} 1 \\ 1 \\ 0 \end{bmatrix}$.

Chapter 4 Concepts and solutions of differential equations

4.1 Concepts

Definition 4.1.1. A differential equation[1] is an equation that involves unknown functions and their derivatives.

If the functions are real functions of one real variable, the derivatives occurring are ordinary derivatives, and the equation is called an ordinary differential equation[2]. If the functions are real functions of more than one real variable, the derivatives occurring are partial derivatives, and the equation is called a partial differential equation[3]. When we refer to an equation as a differential equation, we shall mean an ordinary differential equation. Until we state otherwise, we shall restrict ourselves to differential equation involving a single unknown function. The following are examples of differential equations:

$$\frac{dy}{dx}=x \tag{4.1.1}$$

$$\frac{d^2y}{dx^2}=\sin x \tag{4.1.2}$$

$$\frac{d^2y}{dt^2}+\left(\frac{dy}{dt}\right)^2+\sin y=0 \tag{4.1.3}$$

---

[1] differential equation 微分方程

[2] ordinary differential equation 常微分方程

[3] partial differential equation 偏微分方程

$$\frac{\partial^2 y}{\partial x^2}=\frac{1}{c^2}\frac{\partial^2 y}{\partial t^2} \tag{4.1.4}$$

Equations (4.1.1)～(4.1.3) involve only ordinary derivatives, and they are called ordinary differential equations. Equation (4.1.4) involves partial derivatives and so it called a partial differential equation

Definition 4.1.2. By a solution of the DE (differential equation)

$$F\left(x, y, \frac{dy}{dx}, \frac{d^2 y}{dx^2}, \cdots, \frac{d^n y}{dx^n}\right)=0 \tag{4.1.5}$$

we mean a real function f, denoted by $y=f(x)$, defined on a set of real numbers, where S is the union of non-overlapping intervals such that

$$F(x, f(x), f'(x), f''(x), \cdots, f^n(x)) \equiv 0$$

for all x in S. If x_0 is a left (right) endpoint of an interval in S, derivatives on the right (left) of f at x_0 are intended. Note: The variables x and y need not appear in equation (4.1.5); however, at least one derivative must appear if equation (4.1.5) is to be termed a DE.

【Example 4.1.1】 The function $y(t)=2\sin t-\frac{1}{3}\cos 2t$ is a solution of the differential equation

$$\frac{d^2 y}{dt^2}+y=\cos 2t.$$

Since

$$\frac{d^2}{dt^2}\left(2\sin t-\frac{1}{3}\cos 2t\right)+\left(2\sin t-\frac{1}{3}\cos 2t\right)$$
$$=\left(-2\sin t+\frac{4}{3}\cos 2t\right)+2\sin t-\frac{1}{3}\cos 2t=\cos 2t$$

【Example 4.1.2】 The function defined by $y=f(x)=e^{2x}\equiv\exp(2x)$ is a solution of the DE

$$y'-2y=c \text{ on}(-\infty,+\infty).$$

Since $2e^{2x}-2(e^{2x})\equiv 0$ for all x.

In many applications we seek a solution of DE where the domain of the solution is a specified interval. For example, if t denotes time in a DE, a solution is often sought on the interval $t\geqslant 0$. If no domain is specified, we seek a solution (or solutions) having the largest possible domain, consisting of one or more intervals.

【Example 4.1.3】 The function defined by $y=\sqrt{1-x^2}$ is a solution of DE $yy'+x=0$ on $(-1,1)$, since $\sqrt{1-x^2}\left(\frac{-2x}{2\sqrt{1-x^2}}\right)+x\equiv 0$ for all x in $(-1,1)$. Note that y is undefined on $(-\infty,1)\cup(1,+\infty)$ and that y'' is undefined at $x=\pm 1$.

【Example 4. 1. 4】 Find solutions of the DE $y'=x^2$.

Solution. Since $d/dx(-x^{-1}+c)=x^{-2}$ for all $x\neq 0$, then $y=-x^{-1}+c$ defines a family of solutions on $S=(-\infty,0)\cup(0,+\infty)$, where c is an arbitrary constant. Thus, the DE has an infinite number of solutions, each having domain S.

【Example 4. 1. 5】 Find solutions of $y'=\dfrac{x}{\sqrt{x^2-1}}-\dfrac{x}{\sqrt{9-x^2}}$.

Solution. Since

$$\frac{d}{dx}(\sqrt[n]{x^2-1}+\sqrt{9-x^2}+C)=\frac{x}{\sqrt{x^2-1}}-\frac{x}{9-x^2}$$

Then

$$y=\sqrt{x^2-1}+\sqrt{9-x^2}+C$$

defines a family of solutions, each with domain $S=(-3,-1)\cup(1,3)$.

【Example 4. 1. 6】 Find solutions of $y'=\dfrac{3\sqrt{x}}{2}$.

Since
$$\frac{d}{dx}(x^{\frac{3}{2}}+c)=\frac{3\sqrt{x}}{2}$$

Then $y=f(x)=x^{3/2}+C$ defines family of solutions, each having domain $S=[0,+\infty)$. By $f'(0)=0$ we mean the derivative on the right of f at $x=0$.

【Example 4. 1. 7】 The DE $(dy/dx)^2+x^2y^2+1=0$ has no real-valued solution, since the left member is positive for all differentiable real functions of a real variable. This example shows us that there is no guarantee that a given DE has even one solution.

【Example 4. 1. 8】 The DE $(dy/dx)^2+y^2=0$ has the unique solution f with specified domain $(-\infty,+\infty)$ where$y=f(x)\equiv 0$; that is, the DE has one and only one solution. If a solution g existed on $(-\infty,+\infty)$ where $g(x_0)\neq 0$ at some $x=x_0$, the left member of the DE would be positive for $x=x_0$. Note that the solution of this involves no arbitrary constants.

Order. The order❶ of a differential equation is the order of the highest derivative that appears in the equation. Thus equation(4. 1. 1) and example (4. 1. 2)-(4. 1. 7) are first order equation. Equations(4. 1. 2)~(4. 1. 4) and example 4. 1. 1 are second order differential equation.

【Example 4. 1. 9】 The DE $y'''-x^2=0$ is a third -order equation.

More generally, the equation

$$F(x,u(x),u'(x),\cdots,u^{(n)}(x))=0 \qquad (4.1.6)$$

is an ordinary differential equation of the n th order. Equation (4. 1. 6) represents a

---

❶ order 阶

relation between the independent variable x and the values of the function u and its first n derivatives $u', u'', \cdots, u^{(n)}$. It is convenient and customary in differential equations to write y for $u(x)$, with $y', y'', \cdots, y^{(n)}$ standing for $u', u'', \cdots, u^{(n)}$. Thus equation (4.1.6) is written as

$$F(x, y, y', \cdots, y^{(n)})=0 \tag{4.1.7}$$

【Example 4.1.10】

$$y'''+2e^x y''+yy'=x^4 \tag{4.1.8}$$

is a third order differential equation for $y=u(x)$.

Linear and nonlinear[1]. The ordinary differential equation

$$F(x, y, y', \cdots, y^{(n)})=0$$

is said to be linear if F is a linear function of the variables $y, y', \cdots, y^{(n)}$; a similar definition applies to partial differential equation. Thus the general linear ordinary differential equation of order n is

$$a_0(x)y^{(n)}+a_1(x)y^{(n-1)}+\cdots+a_n(x)y=g(x) \tag{4.1.9}$$

An equation that is not of the form of equation (4.1.9) is a nonlinear equation. Thus a linear DE is linear in y and its derivatives.

【Example 4.1.11】 The DE $y''+\cos x=0$ is nonlinear, since it is not linear in y.

【Example 4.1.12】 The DE $y''+yy'+x=0$ is nonlinear, since y, the coefficient of y' denotes an unknown function of x instead of a specific function of x.

Definition 4.1.3. By a general solution of an nth-order DE, we mean a solution containing n essential arbitrary constants. If every solution of a DE can be obtained by assigning particular values to the n arbitrary constants in a general solution is called a complete solution[2]. A solution of the DE that can be obtained from a general solution by assigning particular values to the arbitrary constants is called a singular solution[3].

The term *general solution* is unsatisfactory since solution may or may not be a complete solution. Some authors use "general solution" only when the treating linear DE.

【Example 4.1.13】 The solution given by $y=\sin x+c$ is a general solution of the DE $y'=\cos x$ This solution is also a complete solution, since two functions having' a derivative given by $\cos x$ can differ by at most a constant.

【Example 4.1.14】 The solution given by $y=cx-c^2$, where c is an arbitrary constant, is a general solution of the first-order DE $\left(\frac{dy}{dx}\right)^2-x\,dy/dx+y=0$. This is

---

[1] linear and nonlinear 线性和非线性

[2] complete solution 全解

[3] singular solution 奇解

not a complete solution, since the DE also has the singular given by $y=x^2/4$. The singular solution cannot be obtained from $y=cx-c^2$ by assigning a particular value to c.

Definition 4.1.4. Any solution of a DE that can be obtained from a general solution by assigning values to the essential arbitrary constants is called a *particular solution*.

For example, by setting $c=0$ in example 4.1.13, we obtain the particular solution given by $y=\sin x$ of the DE $y'=\cos x$.

【Example 4.1.15】 Solutions of the DE $y'=2x^{-3}$ are given by $y=c-x^{-2}$, and the domain of each solution is the set $S=(-\infty,0)\cup(0,+\infty)$.

The following example illustrates a partial DE.

【Example 4.1.16】 Show that the function f defined by $z=f(x,t)=(x-4t)^2$ is a solution of the partial DE

$$\frac{\partial^2 z}{\partial t^2}=16\frac{\partial^2 z}{\partial x^2}$$

Solution. From

$$\frac{\partial z}{\partial x}=2(x-4t),\quad \frac{\partial^2 z}{\partial x^2}=2$$

$$\frac{\partial z}{\partial t}=-8(x-4t),\quad \frac{\partial^2 z}{\partial t^2}=32$$

we obtain $32=16(2)$, true for all x and all t. The domain of f is the set of all ordered pairs (x,t) of real numbers. A special type of DE is classified according to degree.

Definition 4.1.5. If the DE

$$F(x,y,y',y'',\cdots,y^{(n)})=0 \tag{4.1.10}$$

can be expressed as a polynomial in $y',y'',\cdots,y^{(n)}$ the exponent of the highest-order derivative is called the degree[1] of the DE.

【Example 4.1.17】 The differential equations

$$y'''-x^2=0,\ y''-y'^2+x=0,\ y'''-3y'+2y-e^x=0$$

$$x^2y''+xy'+x^2=0 \quad \text{and} \quad \frac{\partial^2 z}{\partial x^2}+\frac{\partial^2 z}{\partial y^2}$$

are first-degree equations. The DE

$$(y')^2-xy'+y=0 \quad \text{and} \quad (y')^2-xy^3=0$$

---

[1] degree 阶

are second-degree equations.

The DE $y''=\pm\sqrt{1+y'}$ is also a second-degree equation since it can be written in the form $(y'')^2-y'=1$ The DE $y''-\ln y=0$ and $y'-\cos y=0$ have no degree since neither can be written in the form equation (4.1.10).

We make the usual assumption that the DE

$$F(x,y,y',y'',\cdots,y^{(n)})=0 \tag{4.1.11}$$

can be solved for the highest-order derivative appearing; that is, that it can be written in the normal from

$$y^{(n)}=G(x,y,y',y'',\cdots,y^{(n-1)}) \tag{4.1.12}$$

which is a DE of degree one.

The reason for this assumption is that DE (4.1.11) may have the same solutions as two or more DE of the form equation (4.1.12). The assumption enables us to treat one DE at a time.

For example, the DE $(y')^2-4x^2=0$ has the same solutions as the two DE

$$y'=2x,\ y'=-2x$$

having solutions given by $y=f(x)=x^2+c_1$ and $y=g(x)=-x^2+c_2$. It is easy to verify that f and g define solutions of $(y')^2-4x^2=0$ on $(-\infty,+\infty)$. Under our assumption, we would treat $y'=2x$ and $y'=-2x$ separately.

4.2 Existence and uniqueness of solutions

The basic problem of the quantitative theory of differential equations is to find all solutions of a given DE. This is a formidable task, since most of DE do not have solutions that can be expressed as combinations of the trigonometric, inverse trigonometric, exponential, logarithmic, and algebraic functions encountered in the calculus.

An important problem in the qualitative theory of DE consists of proving that under certain conditions, solutions of a given DE exist. Theorems of this nature, termed *existence theorems*, were introduces by the French mathematician Augustin-Louis Cauchy (1789-1875) in the 1820s. His investigations marked the beginning of the modern theory of DE. Existence theorems were later studied by such prominent mathematicians as Lipschitz (1832-1903), Picard (1856-1912). Peano (1858-1932), and Liapounoff (1857-1918). Another important qualitative problem consists of proving that a particular solution satisfying certain given conditions is unique; that is, that the solution is the only solution. Theorems of this kind are termed *uniqueness theorems*. When a DE is used a mathematical model to describe a physical situation, it is gratifying to know in advance that the DE has a unique solution satisfying the given conditions of the physical problem. If the DE did not have a unique solution, its appropriateness as a model would be in serious question.

Another aspect of the qualitative theory of DE involves the study of properties of solutions without actually solving the given DE. For example, we can investigate the boundedness of solutions, the periodicity of solutions, the existence of asymptotes of the solution curves, and we can try to construct approximate graphs of solutions. We can also use numerical methods to approximate values of solutions of DE to any required degree of accuracy. Computing machines are very useful in this type of attack.

We can establish an existence and uniqueness theorem for DE of the form $y'=f(x)$.

Theorem 4.2.1. Let f be continuous on an interval I containing a and let k be a given constant. Then the DE $y'=f(x)$ has a unique solution φ on I such that $\varphi(a)=k$.

Proof. Since f is continuous on I, the indefinite integral F of f, given by $F(x)=\int_a^x f(t)\mathrm{d}t$, exists for all $x \in I$.

By the first form of the fundamental theorem of calculus $F'(x)=f(x)$, on I. Let φ be the function defined on I by

$$y=\varphi(x)=k+\int_a^x f(t)\mathrm{d}t$$

Since $\varphi'(x)=f(x)$ on I and $\varphi(x)=k$, therefore φ is a solution guaranteed by the theorem. This complete the existence portion of the theorem.

To prove that φ is the only solution of $y'=f(x)$, let ψ denote an arbitrary solution on I of $y'=f(x)$ for which $\psi(a)=k$. Since φ and ψ are primitives, or antiderivatives[1] of f on I, $\psi(x)-\varphi(x)=c$ on I.

And hence

$$\psi(a)-\varphi(a)=k-k=0=c$$

Thus, $\psi(x)=\varphi(x)$ on I; that is, no solution different from φ exists. In other words, the solution φ is unique.

【Example 4.2.1】 Solve the DE

$$y'=f(x)=3x^2+2x+1 \tag{4.2.1}$$

Solution. Since the antiderivative of a sum is the sum of the antiderivatives, and since any two antiderivatives of f differ by at most a constant, a complete solution is given by $y=x^3+x^2-x+c$.

Specific or particular solutions can be obtained by assigning proper values to c and every solution has domain $(-\infty,+\infty)$.

---

[1] antiderivatives 不定积分

【**Example 4.2.2**】 Solve the DE $y' = e^{-x^2}$, given that $y = 1$ when $x = 0$.

Solution. By Theorem 4.2.1 the unique solution is given by

$$y = \varphi(x) = 1 + \int_0^x e^{-t^2} dt$$

Since the function f is given by $f(x) = e^{-x^2}$ is continuous on $(-\infty, +\infty)$, the solution φ has domain $(-\infty, +\infty)$. The DE is regarded as "solved", even though we cannot readily evaluate $\varphi(x_0)$ for a given x_0. We could estimate $\varphi(x_0)$ by means of an approximating technique. The approximate integration can be performed on a computer and the results presented in a table. A DE is customarily considered solved when its solution has been reduced to one or more integrations[1] (sometimes referred to as quadratures), and the integrals involves are known to exists. This explains why a solution of a DE is sometimes referred to as an integral of the DE.

If we employ the extensively tabulated error function define by

$$\text{erf}x = \frac{2}{\sqrt{\pi}} \int_0^x e^{-t^2} dt$$

we can write our solution as $\quad y = 1 + \frac{\sqrt{\pi}}{2} \text{erf}x$

The important point is not that erf is a known function, but rather that $\int_0^x e^{-t^2} dt$ defines a function.

【**Example 4.2.3**】 Show that any differentiable function[2] implicitly on an interval I by the equation

$$\ln(x^2 + y^2) = \tan^{-1}\frac{y}{x} + c \tag{4.2.2}$$

is a solution on I of the DE

$$\frac{dy}{dx} = \frac{2x + y}{x - 2y}$$

Solution. Differentiating implicitly, we obtain

$$\frac{2x + 2yy'}{x^2 + y^2} = \frac{1}{1 + y^2 x^{-2}} \cdot \frac{xy' - y}{x^2} = \frac{xy' - y}{x^2 + y^2}$$

Multiplying both sides by $x^2 + y^2$ where $x^2 + y^2 \neq 0$, and solving for y' yields

$$y' = \frac{dy}{dx} = \frac{2x + y}{x - 2y}$$

The given DE is regarded as "solved" with solution (or solutions) defined implicitly

---

[1] integrations 积分

[2] differentiable function 可微函数

by equation (4.2.2).

The equation of whether or not an implicit equation defines one or more differentiable function belongs to the theory of implicit functions.

We also regard a DE as solved if we can express a solution (or solutions) parametrically by equations of the form $y=\varphi(t)$, $x=\psi(t)$.

【Example 4.2.4】 Show that the parametric equations

$$y=5-e^{-t};\ x=2e^{t},\quad -\infty<t<+\infty$$

define a solution of the DE

$$\frac{dy}{dx}=\frac{dy/dt}{dx/dt}=\frac{e^{-t}}{2e^{t}}=\frac{5-(5-e^{-t})}{2e^{t}}=\frac{5-y}{x}$$

It is interesting to note that DE often given rise to, or define, new functions. For example, the DE

$$x^2y''+xy'+(x^2-p^2)y=0 \tag{4.2.3}$$

where p is a non-negative constant, it is known as *Bessels equation*, after the German astronomer Friedrich Wilhelm Bessel (1784-1846), who encountered it in a problem on planetary motion. The same equation arise in problem involving heat flow in cylinders, prop agation of electric currents in cylindrircal conductors, vibration of membranes, vibration of chains, and in numerous other important investigations. This is one of the many instances in which the same mathematical model is appropriate for describing several different physical situations. The solutions of equation (4.2.3) have been investigated indetail, have been extensively tabulated, and are termed *Bessel functions*.

Another example is furnished by the DE $d^2y/d\theta^2+y=0$, from which it is possible to develop analytic trigonometry. The solution f or which $y=0$ when $\theta=0$ is given by $y=\sin\theta$, the solution for which $y=1$ when $\theta=0$ is given by $y=\cos\theta$, and a complete solution is given by $y=c_1\sin\theta+c_2\cos\theta$.

A complete solution is also given by $y=A\sin(\theta+\varphi)$, where

$$A=\sqrt{c_1^2+c_2^2},\quad \varphi=\tan^{-1}\left(\frac{c_2}{c_1}\right)$$

【Example 4.2.5】 Solutions of $y'=2x^{-3}$ are given by $y=c-x^{-2}$, each solution having domain $S=(-\infty,0)\cup(0,+\infty)$. By Theorem 4.2.1, $y=c-x^{-2}$ defines a complete solution on $(-\infty,0)$ and also on $(0,+\infty)$. On $S=(-\infty,0)\cup(0,+\infty)$, however, $y=c-x^{-2}$ does not define a complete solution. The solution g given by

$$y=g(x)=\begin{cases}1-x^{-2},x\in(-\infty,0)\\ 2-x^{-2},x\in(0,+\infty)\end{cases}$$

is a solution on S and yet it cannot be obtained by assigning a single value to c. This difficulty does not arise in an application in which domain of a solution consists of a single interval.

4.3 First-order linear differential equations

We begin by studying first-order differential equations and we will assume that our equation is, or can be put, in the form

$$\frac{\mathrm{d}y}{\mathrm{d}t}=f(t,y) \tag{4.3.1}$$

The problem before us is this : given $f(t,y)$ find all functions $y(t)$ which satisfy the differential equation (4.3.1). We approach this problem in the following manner. A fundamental principle of mathematics is that the way to solve a new problem is to reduce it, in some manner, to a problem that we have already solved. In practice this usually entails successively simplifying the problem until it resembles one we have already solved. Since we are presently in the business of solving differential equations, it is advisable for us to take inventory and list all the differential equations we can solve. If we assume that our mathematical background consists of just elementary calculus then the vary sad fact is that the only first-order differential equation we can solve at present is

$$\frac{\mathrm{d}y}{\mathrm{d}t}=g(t) \tag{4.3.2}$$

where g is any integrable function of time. To solve equation (4.3.2) simply integrate both sides with respect to t, which yields

$$y(t)=\int g(t)\mathrm{d}t+c$$

Here c is an arbitrary constant of integration, and by $g(t)\mathrm{d}t$ we mean an anti-derivative of g, that is, a function whose derivative is g. Thus, to solve any other differential reduce it to the equation we must somehow reduce it to the form equation (4.3.2). But this is impossible to do in most cases. Hence, we will not be able, without the aid of a computer, to solve most differential equations. It stands to reason, therefore, that to find those differential equations that we can solve, we should start with very simple equations and not ones like

$$\frac{\mathrm{d}y}{\mathrm{d}t}=\mathrm{e}^{\sin(t-37\sqrt{|y|})}$$

(which incidentally, cannot be solved exactly). Experience has taught us that the "simplest" equation are those which are *linear* in the dependent variables.

Definition 4.3.1. The general first-order linear equation[1] is

---

[1] first-order linear equation 一阶线性方程

$$\frac{dy}{dt}+a(t)y=b(t) \tag{4.3.3}$$

Unless otherwise stated, the functions $a(t)$ and $b(t)$ are assumed to be continuous function of time. We single out this equation and call it linear because the dependent variable y appears by itself, that is, no terms such as e^{-y}, y^3 or $\sin y$ etc. appear in the equation. For example $dy/dt=y^2+\sin t$ and $dy/dt=\cos y+t$ are both *non-linear* equations because of the y^2 and $\cos y$ terms respectively.

Now it is not immediately approach how to solve equation (4.3.3). Thus, we simplify it even further by setting $b(t)=0$.

Definition 4.3.2. The equation

$$\frac{dy}{dt}+a(t)y=0 \tag{4.3.4}$$

is called the homogeneous first-order linear differential equation[1], and equation (4.3.3) is called the *non-homogeneous* first-order linear differential equation for $b(t)$ not identically zero.

Fortunately, the homogeneous equation (4.3.4) can be solved quite easily. First, divide both sides of the equation by y and rewrite it in the form

$$\frac{\frac{dy}{dt}}{y}=-a(t)$$

Second, observe that

$$\frac{\frac{dy}{dt}}{y}=\frac{d}{dt}\ln|y(t)|$$

where by $\ln|y(t)|$ we mean the natural logarithm of $|y(t)|$. Hence equation (4.3.4) can be written in the form

$$\frac{d}{dt}\ln|y(t)|=-a(t) \tag{4.3.5}$$

But this is equation (4.3.2) "essentially" since we can integrate both sides of equation (4.3.5) to obtain that

$$\ln|y(t)|=-\int a(t)dt+c_1$$

where c_1 is an arbitrary constant of integration.

Taking exponentials of both sides yields

$$|y(t)|=\exp\left(-\int a(t)dt+c_1\right)=c\exp\left(-\int a(t)dt\right)$$

[1] homogeneous first-order linear differential equation 齐次一阶线性微分方程

or

$$\left| y(t)\exp\left(\int a(t)\mathrm{d}t\right) \right| = c \tag{4.3.6}$$

Now, $y(t)\exp\int a(t)\mathrm{d}t$ is a continuous function of time and equation (4.3.6) states that its absolute value is constant. But if the absolute value of a continuous function $g(t)$ is constant then g itself must be constant. To prove this observe that if g is not constant, then there exist two different time t_1 and t_2 for which $g(t_1)=c$and $g(t_2)=-c$. By the intermediate value theorem of calculus g must achieve all value between $-c$ and $+c$ which is impossible if $|g(t)|=c$. Hence, we obtain the equation

$$y(t)\exp\left(\int a(t)\mathrm{d}t\right) = c$$

or

$$y(t) = c\exp\left(\int -a(t)\mathrm{d}t\right) \tag{4.3.7}$$

Equation (4.3.7) is said to be the *general solution* of the homogeneous equation since every solution of equation (4.3.4) must be of this form. Observe that an arbitrary constant c appears in equation (4.3.7). This should not be too surprising. Indeed, we will always expect an arbitrary constant to appear in the general solution of any first-order differential equation. To wit, if we give $\mathrm{d}y/\mathrm{d}t$ and we want to recover $y(t)$, then we must perform an integration, and this, of necessity, yields an arbitrary constant. Observe also that equation (4.3.4) has infinitely many solutions; for each value of c we obtain a distinct solution $y(t)$.

【Example 4.3.1】 Find the general solution of the equation $(\mathrm{d}y/\mathrm{d}t)+2ty=0$.

Solution. Here $a(t)=2t$ so that

$$y(t) = \exp\left(-\int 2t\,\mathrm{d}t\right) = c\mathrm{e}^{-t^2}$$

【Example 4.3.2】 Deteremine the behavior, as $t\to\infty$, of all solution of the equation $(\mathrm{d}y/\mathrm{d}t)+ay=0$, a is a constant.

Solution. The general solution is

$$y(t) = c\exp\left(-\int a\,\mathrm{d}t\right) = c\mathrm{e}^{-at}$$

Hence if $a<0$, all solutions, with the exception of $y=0$, approach infinity, and if $a>0$, all solutions approach zero as $t\to\infty$. In applications, we are usually not interested in all solutions of equation (4.3.4). Rather, we are looking for the specific solution $y(t)$ which at some initial time t_0 has the value y_0.

Thus, we want to determine a function $y(t)$ such that

$$\frac{\mathrm{d}y}{\mathrm{d}t}+a(t)y=0,\ y(t_0)=y_0 \tag{4.3.8}$$

Equation (4.3.8) is referred to as an initial-value problem for the obvious reason that of the totality of all solutions of the differential equation, we are looking for one solution which initially (at time t_0) has the value y_0. To find this solution we integrate both sides of equation (4.3.5) between t_0 and t. Thus

$$\int_{t_0}^{t} \frac{d}{ds}\ln|y(s)|\,ds = -\int_{t_0}^{t} a(s)ds$$

and, therefore

$$\ln|y(t)| - \ln|y(t_0)| = \ln\left|\frac{y(t)}{y(t_0)}\right| = -\int_{t_0}^{t} a(s)ds$$

Taking exponentials of both sides of this equation we obtain that

$$\left|\frac{y(t)}{y(t_0)}\right| = \exp\left(-\int_{t_0}^{t} a(s)ds\right)$$

or

$$\left|\frac{y(t)}{y(t_0)}\exp\left(\int_{t_0}^{t} a(s)ds\right)\right| = 1$$

The function inside the absolute value sign is a continuous function of time. Thus, by the argument given previously, it is identically $+1$ or identically -1. To determine which one it is, evaluate it at the point t_0; since

$$\frac{y(t_0)}{y(t_0)}\exp\left(\int_{t_0}^{t_0} a(s)ds\right) = 1$$

we see that

$$\frac{y(t)}{y(t_0)}\exp\left(\int_{t_0}^{t} a(s)ds\right) = 1$$

Hence

$$y(t) = y(t_0)\exp\left(\int_{t_0}^{t} a(s)ds\right) = y_0\exp\left(\int_{t_0}^{t} a(s)ds\right)$$

【Example 4.3.3】 Find the solution of the initial-value problem

$$\frac{dy}{dt} + (\sin t)y = 0,\quad y(0) = \frac{3}{2}$$

Solution. Here $a(t) = \sin t$ so that

$$y(t) = \frac{3}{2}\exp\left(\int_0^t \sin s\,ds\right) = \frac{3}{2}e^{\cos t - 1}$$

【Example 4.3.4】 Find the solution of the initial-value problem

$$\frac{dy}{dt} + e^{t^2}y = 0,\quad y(1) = 2$$

Solution. Here $a(t) = e^{t^2}$ so that

$$y(t) = 2\exp\left(-\int_1^t e^{s^2}ds\right)$$

Now, at first glance this problem would seem to present a very serious difficulty in that we cannot integrate the function e^{s^2} directly. However, this solution is equally as valid and equally as useful as the solution to example 4. 3. 3. The reason for this is twofold. First, there are very simple numerical schemes to evaluate the above integral to any degree of accuracy with the explicitly, we still cannot evaluate it at any time t without the aid of a table of trigonometric functions and some sort of calculating aid, such as a slide rule, electronic calculator or digital computer.

We return now to the non-homogeneous equation

$$\frac{dy}{dt}+a(t)y=b(t).$$

It should be clear from our analysis of the homogeneous equation that the way to solve the non-homogeneous equation is to express it in the form

$$\frac{d}{dt}(\text{“}something\text{”})=b(t)$$

and then to integrate both side to solve for “something”. However, the express $(dy/dt)+a(t)y$ does appear to be the derivative of some simple expression. The next logical step in our analysis therefore should be the following: Can we make the left hand side of the equation to be d/dt of “something”? More precisely, we can multiply both sides of equation (4. 3. 3) by any continuous function $\mu(t)$ to obtain the equivalent equation

$$\mu(t)\frac{dy}{dt}+a(t)\mu(t)y=\mu(t)b(t) \tag{4. 3. 9}$$

By equivalent equations we mean that every solution of equation (4. 3. 9) is a solution of equtaion (4. 3. 3) and viceversa. Thus, can we choose $\mu(t)$ so that $\mu(t)(dy/dt)+a(t)\mu(t)y$ is the derivative of some simple expression? The answer to this equation is yes, and is obtained by observing that

$$\frac{d}{dt}\mu(t)y=\mu(t)\frac{dy}{dt}+\frac{d\mu}{dt}y$$

Hence, $\mu(t)(dy/dt)=a(t)\mu(t)y$ will be equal to the derivative of $\mu(t)y$ if and only if $d\mu(t)/dt=a(t)\mu(t)$. But this is a first-order linear homogeneous equation for $\mu(t)$, i. e. $(d\mu/dt)-a(t)\mu=0$ which we already know how to solve, and since we only need one such function $\mu(t)$ we set the constant c in equation (4. 3. 7) equal to one and take

$$\mu(t)=\exp\left(\int a(t)dt\right)$$

For this $\mu(t)$, equation (4. 3. 9) can be written as

$$\frac{d}{dt}\mu(t)y=\mu(t)b(t) \tag{4. 3. 10}$$

To obtain the general solution of the non homogeneous equation (4. 3. 3), that is, to find all solutions of non-homogeneous equation, we take the indefinite integral (anti-derivative) of both sides of equation (4. 3. 10) which yields

$$\mu(t)y=\int\mu(t)b(t)\mathrm{d}t+c$$

or

$$y=\frac{1}{\mu(t)}\left(\int\mu(t)b(t)\mathrm{d}t+c\right)=\exp\left(-\int a(t)\right)\left(\int\mu(t)b(t)\mathrm{d}t+c\right) \quad (4.3.11)$$

Alternately, if we interested in the specific solution (4. 3. 3) satisfying the initial condition $y(t_0)=y_0$, that is, if we want to solve the initial- value problem

$$\frac{\mathrm{d}y}{\mathrm{d}t}+a(t)y=b(t), \quad y(t_0)=y_0$$

then we can take the definite integral of both sides of (4. 3. 10) between t_0 and t to obtain that

$$\mu(t)y-\mu(t_0)y_0=\int_{t_0}^{t}\mu(s)b(s)\mathrm{d}s$$

or

$$y=\frac{1}{\mu(t)}\left(\mu(t_0)y_0+\int_{t_0}^{t}\mu(s)b(s)\mathrm{d}s\right).$$

Remark 4. 3. 1. Notice how we used our knowledge of the solution of the homogeneous equation to find the function $\mu(t)$ which enables us to solve the non-homogeneous equation, This is an excellent illustration of how we use our knowledge of the solution of a simple problem to solve a harder problem.

Remark 4. 3. 2. The function $\mu(t)=\exp\left(\int a(t)\mathrm{d}t\right)$ is called integrating factor for the non-homogeneous equation since after multiplying both sides by $\mu(t)$ we can immediately integrate the equation to find all solutions.

Remark 4. 3. 3. The reader should not memorize formula (4. 3. 11). Rather, we will solve all non-homogeneous equations by first multiplying both sides by $\mu(t)$, by writing the new left-hand side as the derivative of $\mu(t)y(t)$, and then by integrating both sides of the equation.

Remark 4. 3. 4. An alternative way of solving the initial-value problem

$$(\mathrm{d}y/\mathrm{d}t)+a(t)y=b(t), \ y(t_0)=y_0$$

is to find the general solution (4. 3. 11) of equation (4. 3. 3) and then use the initial condition $\mu(t)b(t)$ cannot be integrated directly, though, then we must take the definite integral of equation (4. 3. 10) to obtain equation (4. 3. 11), and this equation is then approximated numerically.

【Example 4. 3. 5】 Find the general solution of the equation $(\mathrm{d}y/\mathrm{d}t)-2ty=t$.

Solution. Here $a(t)=-2t$ so that

$$\mu(t)=\exp\left(\int a(t)\,dt\right)=\exp\left(-\int 2t\,dt\right)=e^{-t^2}$$

Multiplying both sides of the equation by $\mu(t)$ we obtain the equivalent equation.

$$e^{-t^2}\left(\frac{dy}{dt}-2ty\right)=te^{-t^2} \text{ or } \frac{d}{dt}e^{-t^2}y=te^{-t^2}$$

Hence,

$$e^{-t^2}y=\int te^{-t^2}\,dt+c=\frac{-e^{-t^2}}{2}+c$$

and

$$y(t)=-\frac{1}{2}+ce^{t^2}$$

【Examp 4. 3. 6】 Find the solution of the initial-value problem

$$\frac{dy}{dt}+2ty=t, \qquad y(1)=2$$

Solution. Here $a(2)=2t$ so that

$$\mu(t)=\exp\left(\int a(t)\,dt\right)=\exp\left(\int 2t\,dt\right)=e^{t^2}$$

Multiplying both sides of the equation by $\mu(t)$ we obtain the equivalent equation

$$e^{t^2}\left(\frac{dy}{dt}+2ty\right)=te^{t^2}$$

or

$$\frac{d}{dt}e^{t^2}y=te^{t^2}$$

Hence,

$$\int_1^t \frac{d}{ds}e^{s^2}y(s)\,ds=\int_1^t se^{s^2}\,ds$$

so that

$$e^{s^2}y(s)\Big|_1^t=\frac{e^{s^2}}{2}\Big|_1^t$$

Consequently,

$$e^{t^2}y-2e=\frac{e^{t^2}}{2}-\frac{e}{2}$$

and

$$y=\frac{1}{2}+\frac{3e}{2}e^{-t^2}=\frac{1}{2}+\frac{3}{2}e^{1-t^2}$$

4. 4 Exact equation and separation of variables

We consider first-order DE of the form

$$M(x,y)+N(x,y)\frac{dy}{dx}=0 \tag{4. 4. 1}$$

Where M and N denote continuous functions of x and y possessing continuous first partial derivatives in a region R of the xy plane. We also assume that $N(x,y)\neq 0$ in R, thereby ensuring that $\mathrm{d}y/\mathrm{d}x$ is defines at every point of R.

$$M(x,y)\mathrm{d}x+N(x,y)\mathrm{d}y=0 \tag{4.4.2}$$

The left member of equation (4.4.2) is called a differential form[❶] in two variables. An advantage of form equation (4.4.2) is that it enables us to regard either y or x as an element of the range of the unknown function or functions. It is sometimes convenient to assume that $M(x,y)\neq 0$ in R and to seek solutions of

$$M(x,y)\frac{\mathrm{d}y}{\mathrm{d}x}+N(x,y)=0 \tag{4.4.3}$$

Given by equations of the form $x=g(y)$. Solutions of equation (4.4.1) are then obtained by solving $x=g(y)$ for y in the terms of x.

Definition equation (4.4.2). is said to be exact region R if and only if there exists a function, denoted by $u=u(x,y)$, such that at every point of R

$$\frac{\partial u}{\partial x}=M,\ \frac{\partial u}{\partial y}=N$$

Since

$$\mathrm{d}u=\frac{\partial u}{\partial x}\mathrm{d}x+\frac{\partial u}{\partial y}\mathrm{d}y$$

Equation (4.4.2) is exact if and only if the left member denotes the total (or exact) differential of a function given by $u=u(x,y)$. the reason the term exact is used.

【Example 4.4.1】 The equation

$$y\mathrm{d}x+x\mathrm{d}y=0$$

Is exact, since the left number is seen by inspection to be differential of the function given by $u=xy$.

The equation

$$(2xy+3y^2)\mathrm{d}x+(x^2+9xy^2)\mathrm{d}y=0$$

is also exact, although the truth of this assertion is not evident by inspection. The following theorem states a necessary for an equation to be exact.

Theorem 4.4.1. If $M(x,y)\mathrm{d}x+N(x,y)\mathrm{d}y=0$ is exact in R, then

$$\frac{\partial M}{\partial y}=\frac{\partial N}{\partial x}\ \text{in}\ R$$

Proof. Since the equation is exact in R, there exists a function given by $u=u(x,y)$ such that $\partial u/\partial x=M$ and $\partial u/\partial y=N$ in R. Therefore,

$$\frac{\partial M}{\partial y}=\frac{\partial^2 u}{\partial y\partial x}\quad \text{and}\quad \frac{\partial N}{\partial x}=\frac{\partial^2 u}{\partial x\partial y}$$

---

❶ differential form 微分形式

Since we are assuming that $\partial M/\partial y$ and $\partial N/\partial x$ are continuous in R, a well-known theorem of calculus guarantees that the mixed partials $\partial^2 u/\partial y\partial x$ and $\partial^2 u/\partial x\partial y$ are equal in R.

To show that an equation of the form equation (4.4.2) is not exact in R, we use the contrapositive of Theorem 4.4.1. The contrapositive of a theorem is obtained by negating both the hypothesis and the conclusion of the theorem, and then interchanging them. A theorem and its contrapositive are both true or both false. Hence Theorem 4.4.2 is equivalent to Theorem 4.4.1.

Theorem 4.4.2 If there is at least one point in R at which $\partial M/\partial y \neq \partial N/\partial x$, then the expression $M(x,y)\mathrm{d}x + N(x,y)\mathrm{d}y$ is not exact in R. (Contrapositive of Theorem 4.4.1)

【Example 4.4.2】 There is no region in which $x\cos y\mathrm{d}x + y\cos x\mathrm{d}y = 0$ is exact since in an arbitrary region R of the xy plane,

$$\frac{\partial}{\partial y}(x\cos y) = -x\sin y \neq \frac{\partial}{\partial x}(y\cos x) = -y\sin x$$

The converse of Theorem 4.4.1, if true, would furnish a test for exactness. That is, $\partial M/\partial y = \partial N/\partial x$ in R. This converse can be established provided that suitable restrictions are placed on R. For simplicity, we assume R to be the interior of a rectangle with sides parallel to the x and y axes.

Theorem 4.4.3 If $\partial M/\partial y = \partial N/\partial x$ in the interior of a rectangle with sides parallel to the x and y axes, then the equation $M(x,y)\mathrm{d}x + N(x,y)\mathrm{d}y = 0$ is exact in the interior of the rectangle.

$$\frac{\partial u}{\partial x} = \frac{\mathrm{d}}{\mathrm{d}x}\int_a^x M(s,b)\mathrm{d}s + \frac{\partial}{\partial x}\int_b^y N(x,t)\mathrm{d}t$$

Proof. Let (a,b) be any convenient fixed point in the rectangle R and let u be the function defined by

$$u(x,y) = \int_a^x M(s,b)\mathrm{d}s + \int_b^y N(x,t)\mathrm{d}t \tag{4.4.4}$$

Where (x,y) is an arbitrary point in R. Then

Since N and $\partial N/\partial x$ are continuous in a rectangle $b \leqslant t \leqslant y, x_1 \leqslant x \leqslant x_2$,

$$\frac{\partial}{\partial x}\int_b^y N(x,t)\mathrm{d}t = \int_b^y \frac{\partial N(x,t)}{\partial x}\mathrm{d}t$$

This result, known as Leibniz's rule for differentiating under the integral sign. Hence, since $\partial N/\partial x = \partial M/\partial y$,

$$\frac{\partial u}{\partial x} = \frac{\mathrm{d}}{\mathrm{d}x}\int_a^x M(s,b)\mathrm{d}s + \int_b^y \frac{\partial N(x,t)}{\partial x}\mathrm{d}t$$

$$= M(x,b) + \int_b^y \frac{\partial M(x,t)}{\partial t}\mathrm{d}t$$

$$=M(x,b)+[M(x,t)]_b^y$$
$$=M(x,b)+M(x,y)-M(x,b)=M(x,y)$$

Also,

$$\frac{\partial u}{\partial y}=\frac{\mathrm{d}}{\mathrm{d}y}\int_a^x M(s,b)\mathrm{d}s+\frac{\partial}{\partial y}\int_b^y N(x,t)\mathrm{d}t$$
$$=0+N(x,y)=N(x,y)$$

This completes the proof of exactness.

In $\int_a^x M(s,b)\mathrm{d}s$, the integrand is the function of x assumed by $M(x,y)$ on the line segment between (a,b) and (x,b), y having the constant value b. In $\int_b^y N(x,t)\mathrm{d}t$, the integrant is the function of y assumed by $N(x,y)$.

On the line segment between (x,b) and (x,y), with x regarded as fixed during the integration. Students familiar with line integrals will recognize the right member of equation (4. 4. 4) as the line integral $\int_C M\mathrm{d}x+N\mathrm{d}y$ taken over the path C, where C is the "elbow path," consisting of the line segment from (a,b) to (x,b) plus the line segment from (x,b) to (x,y). Line integrals have very important mathematical and physical applications and interpretations.

【Example 4. 4. 3】 The equation $(2xy+3y^3)\mathrm{d}x+(x^2+9xy^2)\mathrm{d}y=0$ is exact in the entire xy plane, since for all values of x and y,

$$\frac{\partial}{\partial y}(2xy+3y^3)=2x+9y^2=\frac{\partial}{\partial x}(x^2+9xy^2)$$

【Example 4. 4. 4】 The equation $2x\ln y\mathrm{d}x+x^2y^{-1}\mathrm{d}y=0$ is exact in any rectangle for which $y>0$, since

$$\frac{\partial}{\partial y}(2x\ln y)=2xy^{-1}=\frac{\partial}{\partial x}(x^2y^{-1})$$

Let $M(x,y)\mathrm{d}x+N(x,y)\mathrm{d}y=0$ be exact. Then there exists a function u, which we denote by $u=u(x,y)$, such that $\partial u/\partial x=M$ and $\partial u/\partial y=N$. Solutions of $M(x,y)\mathrm{d}x+N(x,y)\mathrm{d}y=0$ are defined implicitly by $u(x,y)=c$, where c is a constant. To see this, let $y=f(x)$ denote any differentiable function defined by $u(x,y)=c$. By the chain rule,

$$\frac{\partial u}{\partial x}+\frac{\partial u}{\partial y}\frac{\mathrm{d}y}{\mathrm{d}x}=0$$

$$\frac{\partial u}{\partial x}\mathrm{d}x+\frac{\partial u}{\partial y}\mathrm{d}y=M(x,y)\mathrm{d}x+N(x,y)\mathrm{d}y=0$$

And hence $y=f(x)$ defines a solution of equation (4. 4. 1) or equation (4. 4. 2). To find a solution for which $y_0=f(x_0)$, one determines the constant c from $u(x_0,$

$y_0)=c$.

After applying Theorem 4.4.3 to determine that $M\mathrm{d}x+N\mathrm{d}y=0$ is exact, we still have the problem of obtaining the equation $u(x,y)=c$. The simplest method is to recognize by inspection that $M\mathrm{d}x+N\mathrm{d}y$ is an exact differential.

【Example 4.4.5】 The equation $2xy\mathrm{d}x+x^2\mathrm{d}y=0$ is exact since $(\partial/\partial y)(2xy)=2x=(\partial/\partial x)(x^2)$. Since $\mathrm{d}(x^2y)=2xy\mathrm{d}x+x^2\mathrm{d}y$, the solutions are given by $x^2y=c$, or $y=cx^{-2}$, valid on $S=(-\infty,0)\cup(0,+\infty)$.

【Example 4.4.6】 Solve the initial-value problem

$$\frac{3x^2y+1}{y}\mathrm{d}x-\frac{x}{y^2}\mathrm{d}y=0;\quad y(2)=1$$

Solution. The equation is exact since

$$\frac{\partial}{\partial y}\left(\frac{3x^2y+1}{y}\right)=-y^{-2}=\frac{\partial}{\partial x}\left(\frac{-x}{y^2}\right)$$

Write the equation in the form

$$\frac{y\mathrm{d}x-x\mathrm{d}y}{y^2}+3x^2\mathrm{d}x=0$$

And remembering that $\mathrm{d}(x/y)=(y\mathrm{d}x-x\mathrm{d}y)/y^2$, we obtain $x/y+x^3=c$. Setting $x=2$ and $y=1$, we find that $c=10$. Solving for y, we get the required solution given by $y=x/(10-x^3)$, valid on $(0,\sqrt[3]{10})$.

The equation $u(x,y)=c$, defining implicitly the solution of $M\mathrm{d}x+N\mathrm{d}y=0$, can also be obtained directly from equation (4.4.4).

【Example 4.4.7】 Solve $(2xy+3y^3)\mathrm{d}x+(x^2+9xy^2)\mathrm{d}y=0$, shown in Example 4.4.3 to be exact.

Solution. A natural choice for a and b in equation (4.4.4) is $a=b=0$. In this case, however, $N(0,0)=0$, so let us choose $a=1$ and $b=0$. From equation (4.4.4), we obtain

$$u(x,y)=\int_1^x 0\mathrm{d}s+\int_0^y(x^2+9xt^2)\mathrm{d}t$$

$$=k+[x^2t+3xt^3]_0^y=k+x^2y+3xy^3$$

Incorporating the arbitrary constant k into an arbitrary constant c, we obtain $x^2y+3xy^3=c$. It is not easy to solve this equation for y in terms of x.

If equation (4.4.2) can be written in the form of

$$M(x)\mathrm{d}x+N(y)\mathrm{d}y=0 \tag{4.4.5}$$

Where M is a function of x alone and N is a function of y only, the variables x and y in equation (4.4.11) are said to be separable. Equation (4.4.5) is exact, since

$$\frac{\partial}{\partial y}M(x)\equiv 0\equiv\frac{\partial}{\partial x}N(y)$$

Let us assume that $\int M(x)\mathrm{d}x = G(x)+c_1$ and $\int N(y)\mathrm{d}y = H(y)+c_2$.

Then

$$\frac{\mathrm{d}G(x)}{\mathrm{d}x}=M(x) \quad \text{and} \quad \frac{\mathrm{d}H(y)}{\mathrm{d}y}=N(y)$$

And since $M(x)\mathrm{d}x+N(y)\mathrm{d}y$ is the differential of $G(x)+H(y)$, a general solution of equation (4.4.1) is given implicitly by

$$G(x)+H(y)=c \tag{4.4.6}$$

If $y=f(x)$ denotes any differentiable function defined by equation (4.4.6), it follows from the chain rule that f is a solution of equation (4.4.5).

If equation (4.3.2) can be written in the form

$$A(x)B(y)\mathrm{d}x+C(x)D(y)\mathrm{d}y=0 \tag{4.4.7}$$

The variables are then separable, since division by $B(y)C(x)$ yields the equation

$$\frac{A(x)}{C(x)}\mathrm{d}x+\frac{D(y)}{B(y)}\mathrm{d}y=0 \tag{4.4.8}$$

Definition 4.4.1. Two DE are said to be equivalent[1] if and only if they have the same solutions.

The solutions of equation (4.4.7) could be obtained by solving equation (4.4.8) if two DE were equivalent. They may not be, however, since division by $B(y)C(x)$ may introduce discontinuities or extraneous solutions. Such possibilities are investigated by considering the equations $B(y)=0$ and $C(x)=0$.

【Example 4.4.8】 The solutions of $x\mathrm{d}x+y^2\mathrm{d}y=0$ are given implicitly by

$$\frac{x^2}{2}+\frac{y^3}{3}=c$$

【Example 4.4.9】 Solve $1\mathrm{d}x+(y^4+x^2y^4)\mathrm{d}y=0$.

Solution. The given DE is equivalent to the DE

$$\frac{\mathrm{d}x}{1+x^2}+y^4\mathrm{d}y=0 \tag{4.4.9}$$

The solutions are given implicitly by

$$\tan^{-1}x+\frac{y^5}{5}=c$$

【Example 4.4.10】 Solve

$$\frac{\mathrm{d}y}{\mathrm{d}x}=ky, \quad k\neq 0 \tag{4.4.10}$$

---

[1] equivalent 等价

Solution. Dividing the equivalent DE by y, we obtain

$$k\,\mathrm{d}x-\frac{\mathrm{d}y}{y}=0 \tag{4.4.11}$$

If y satisfies equation (4.4.11), then

$$kx-\ln|y|=c,\quad \ln|y|=kx-c$$

and

$$|y|=\mathrm{e}^{kx-c}=\mathrm{e}^{-c}\mathrm{e}^{kx}=P\mathrm{e}^{kx}$$

where P is an arbitrary positive constant.

It can now be shown that $y=\pm P\mathrm{e}^{kx}=B\mathrm{e}^{kx}$, where B is an arbitrary nonzero constant. That is, every differentiable y for which $|y|=P\mathrm{e}^{kx}$ $(P>0)$ is of the form $y=B\mathrm{e}^{kx}$, $B\neq 0$.

Since we divided by y to obtain equation (4.4.11) from equation (4.4.10), we must investigate to see whether $y=0$ defines a solution of equation (4.4.10). This is seen by inspection to be the case, and hence all solutions of equation (4.4.10) must be the form $y=B\mathrm{e}^{kx}$, where B is an arbitrary constant. Conversely, if y is of the form $y=B\mathrm{e}^{kx}$, y satisfies equation (4.4.9) since

$$Bk\mathrm{e}^{kx}=k(B\mathrm{e}^{kx})$$

for all x. Every solution has domain $(-\infty,+\infty)$.

【Example 4.4.11】 Solve the initial-value problem

$$\frac{\mathrm{d}y}{\mathrm{d}x}=ky;\quad k\neq 0,\quad y(0)=5$$

Solution. Setting $x=0$ in $y=B\mathrm{e}^{kx}$, we obtain $B=5$ and $y=5\mathrm{e}^{kx}$.

【Example 4.4.12】 Solve the initial-value problem

$$\frac{\mathrm{d}y}{\mathrm{d}x}=1+y^2;\quad y\left(\frac{\pi}{4}\right)=1$$

Solution.

(1) $\displaystyle\int_1^y \frac{\mathrm{d}u}{1+u^2}=\int_{\pi/4}^x \mathrm{d}t$

(2) $\tan^{-1}y-\tan^{-1}=x-\dfrac{\pi}{4}$

(3) $\tan^{-1}y=x$

(4) $y=\tan x$

(5) The solution has domain $(-\pi/2,\pi/2)$.

4.5 Integrating factors

One can scarcely expect the equation $M\mathrm{d}x+N\mathrm{d}y=0$ to be exact since the requirement $M_y = N_x$ is every special. If the equation is not exact, it is still plausible that there may exists a function given by $\mu = \mu(x, y)$, such that

$$\mu(x, y)M(x, y)\mathrm{d}x + \mu(x, y)N(x, y)\mathrm{d}y = 0$$

is exact. Such a function μ is called an integrating factor[1] of the DE. Euler introduce this useful concept in 1734.

In Section 4.4 we solved the equation

$$A(x)B(y)\mathrm{d}x + C(x)D(y)\mathrm{d}y = 0$$

by considering the equation

$$\frac{A(x)}{C(x)}\mathrm{d}x + \frac{D(y)}{B(y)}\mathrm{d}y = 0.$$

We effectively made the observation that $\mu(x, y) = [C(x)B(y)]^{-1}$ is an integrating factor[1].

Integrating factors are sometimes obtained by inspection; the method depends on the introduction of certain formulas for exact differentials.

【Example 4.5.1】 The equation $-y\mathrm{d}x + x\mathrm{d}y = 0$ is not exact since

$$\frac{\partial}{\partial y}(-y) = -1 \neq \frac{\partial}{\partial x}(x) = +1$$

By recalling the formula $\mathrm{d}(y/x) = (x\mathrm{d}y - y\mathrm{d}x)/x^2$, we note that $\mu(x, y) = x^{-2}$ is an integrating factor. Hence, $y/x = c$, or $y = cx$, defines a general solution. [Note that $y = 0$ defines a solution on $(-\infty, 0) \cup (0, +\infty)$].

By recalling the formula for $\mathrm{d}(-x/y)$, it is easy to show that y^{-2} is also an integrating factor.

【Example 4.5.2】 The equation $(2x+4)\mathrm{d}x + x\mathrm{d}y = 0$ is not exact since

$$(\partial/\partial y)(2y+4) = 2 \neq (\partial/\partial x)(x) = 1$$

If we are clever enough to notice that $\mu(x, y) = x$ is an integrating factor, we obtain $(2xy + 4x)\mathrm{d}x + x^2\mathrm{d}y = 0$, which has solutions given implicitly by

$$x^2 y + 2x^2 = c$$

【Example 4.5.3】 Solve $\quad (x^2 + y^2 - y)\mathrm{d}x + x\mathrm{d}y = 0$

Solution. The equation is not exact since

$$\frac{\partial}{\partial y}(x^2 + y^2 - y) = 2y - 1 \neq \frac{\partial}{\partial x}(x) = 1$$

Writing the equation in the form $(x^2 + y^2)\mathrm{d}x + x\mathrm{d}y - y\mathrm{d}x = 0$, and noting that

$$\mathrm{d}\tan^{-1}\left(\frac{y}{x}\right) = \frac{x\mathrm{d}y - y\mathrm{d}x}{x^2(1 + y^2/x^2)} = \frac{x\mathrm{d}y - y\mathrm{d}x}{x^2 + y^2}$$

We see that $\mu(x, y) = [x^2 + y^2]^{-1}$ is an integrating factor. From

---

[1] integrating factor 积分因子

$$dx+\frac{x\,dy-y\,dx}{x^2+y^2}=0$$

We obtain $x+\tan^{-1}(y/x)=c$.

The following formulas are often exploited in similar fashion:

$$d(xy)=s\,dy+y\,dx$$

$$d(\ln\frac{x}{y})=\frac{y\,dx-x\,dy}{xy}$$

$$d[\frac{1}{2}\ln(x^2+y^2)]=\frac{x\,dx+y\,dy}{x^2+y^2}$$

$$d\sqrt{x^2+y^2}=\frac{x\,dx+y\,dy}{\sqrt{x^2+y^2}}$$

$$d\sqrt{x^2-y^2}=\frac{x\,dx-y\,dy}{\sqrt{x^2-y^2}}$$

In the examples presented, the discovery of an integrating factor required considerable ingenuity. Let us consider the possibility of a more direct determination of μ. If

$$\mu M\,dx+\mu N\,dy=0 \tag{4.5.1}$$

is exact, then

$$\frac{\partial(\mu M)}{\partial y}=\frac{\partial(\mu N)}{\partial x}$$

or

$$\mu\,\frac{\partial(M)}{\partial y}+M\,\frac{\partial\mu}{\partial y}=\mu\,\frac{\partial(N)}{\partial x}+N\,\frac{\partial\mu}{\partial x}$$

and

$$\frac{1}{\mu}\left(N\,\frac{\partial\mu}{\partial x}-M\,\frac{\partial\mu}{\partial y}\right)=\frac{\partial M}{\partial y}-\frac{\partial N}{\partial x} \tag{4.5.2}$$

To find even a particular solution μ of this partial DE is a formidable problem. There are, however, two fairly since special cases of interest.

Suppose that equation (4.5.1) has an integrating factor μ that is a function of x alone. Then

$$\frac{\partial\mu}{\partial x}=\frac{d\mu}{\partial x}\quad\text{and}\quad\frac{\partial\mu}{\partial y}=0$$

and hence equation (4.5.2) reduces to

$$\frac{1}{\mu}\frac{\partial\mu}{\partial x}=\frac{\partial M/\partial y-\partial N/\partial x}{N}. \tag{4.5.3}$$

Since the left member of equation (4.5.3) depends only on x, the same is true of the right member, which we denote by $f(x)$.

From

$$\frac{1}{\mu}\frac{\partial \mu}{\partial x}=f(x) \quad \text{or} \quad \frac{\partial \mu}{\mu}=f(x)\mathrm{d}x$$

We obtain

$$\ln|\mu|=\int f(x)\mathrm{d}x$$

and $\mu = c\mathrm{e}^{\int f(x)\mathrm{d}x}=\mathrm{e}^{\int f(x)\mathrm{d}x}$ for $c=1$.

Conversely, if $(M_y - N_x)/N$ depends on x alone, then $\mu = \mathrm{e}^{\int f(x)\mathrm{d}x}$ satisfies equation (4.5.3) and hence is an integrating factor for $M\mathrm{d}x+N\mathrm{d}y=0$.

【Example 4.5.4】 Solve $\quad -y\mathrm{d}x+x\mathrm{d}y=0$

Solution. From, $(M_y-N_x)/N=(-1-1)/x=-2/x$ we obtain

$$\mu=\mathrm{e}^{\int -2\mathrm{d}x/x}=\mathrm{e}^{-2\ln|x|}=(\mathrm{e}^{\ln|x|})^{-2}=x^{-2}$$

This is the integrating factor we found by inspection in Example 4.5.1. Of course, $-y\mathrm{d}x+x\mathrm{d}y=0$ is easily solved by separating the variables.

4.6 Initial-value and two-point boundary-value

When we solve DE, we often find their solutions in forms from which specific values of these solutions can be determined.

【Example 4.6.1】 Solve the DE $y'=x^2$, given that $y=2$ when $x=1$.

Solution Ⅰ. A general solution, which is also a complete solution, is given by $y=(x^3/3)+c$. Substituting $x=1$ and $y=2$, we obtain $2=\frac{1}{3}+c$, and hence c must equal $\frac{5}{3}$ to satisfy the given condition. Hence the required solution, unique by Theorem 4.2.1, is given by $y=(x^3/3)+\frac{5}{3}$ and has domain $(-\infty,+\infty)$.

Solution Ⅱ. $y=F(x)=c+\int_1^x t^2 \mathrm{d}t$, since the right member is the general antiderivative of x^2. Setting $x=1$, we obtain $2=F(1)=c+0$. Hence,

$$y-2=\int_2^y \mathrm{d}u=\int_1^x t^2\mathrm{d}t \tag{4.6.1}$$

$$y-2=\frac{t^3}{3}\bigg|_1^x=\frac{x^3}{3}-\frac{1}{3}$$

or

$$y=\frac{x^2}{3}+\frac{5}{3}$$

Solution Ⅱ differs from Solution Ⅰ mainly in notation. When the notation of equation (4. 3. 1) is used, the lower limits on the two integral signs are corresponding values of $y$and x, the upper variable limits also correspond, and the variables μ and t are dummy variables.

Variables x and y in Example 4. 6. 1 can be made to satisfy one condition, since our general solution contains one arbitrary constant. A problem of this type is called an initial-value problem, because in many applications one variables is the time, and the specified condition gives the value of the other variable at the initial time $t=0$.

【Example 4. 6. 2】 Solve the second order initial-value problem $y''=6x$, $y=3$, $y'=2$ when $x=0$.

Solution. In $y'=3x^2+c$, we set $x=0$ and $y'=2$ to obtain $y'=3x^2+2$. $\ln y=x^3+2x+k$, we set $x=0$ and $y=3$ to obtain $y=x^3+2x+3$.

【Example 4. 6. 3】 Solve the DE $y''=12x$, given that $y=0$ when $x=0$, and$y=6$ when $x=2$.

Solution. Forming antiderivatives, we obtain $y'=6x^2+c$and

$$y=2x^3+cx+k \tag{4. 6. 2}$$

Applying the given conditions to equation (4. 6. 2) yields $0=k$ and $6=16+2c+k$. The required solution $y=2x^3-5x$ is obtained by substituting $k=0$ and $c=-5$ in equation (4. 6. 2).

In Example 4. 6. 3, variables x and y were made to satisfy two conditions and equation (4. 6. 2) contained two arbitrary constants. This type of problem is called a two-point boundary-value problem[1], since the given conditions usually involve the endpoints of the interval that is of interest and importance in the problem. The solution of many important problems in applied mathematics involves DE. This is because physical laws are usually stated as DE.

【Example 4. 6. 4】 The motion of a particle of mass m moving on the x axis governed by Newton's second law, which states that the force f acting on the particle equals the mass m times the acceleration a of the particle. Since $a=\mathrm{d}^2x/\mathrm{d}t^2$, Newton's second law is expressed as a second-order DE. When we say that the motion is governed by the DE, we mean that the displacement function, denote by $x=h(t)$, is a solution of the DE. If the force f depends on the time t, or the velocity $\mathrm{d}x/\mathrm{d}t$, or the displacement x, or some combination of these there quantities, the DE will have the form

$$F\left(t, x, \frac{\mathrm{d}x}{\mathrm{d}t}, \frac{\mathrm{d}^2x}{\mathrm{d}t^2}\right)=0 \tag{4. 6. 3}$$

which is of the type equation (4. 1. 5), with t replacing x and x replacing y.

---

[1] two-point boundary-value problem 两点边值问题

A general solution of equation (4. 6. 3) would contain two essential arbitrary constants. These constants could be determined from initial conditions of the form

$$t=0, x=x_0;\quad t=0, \frac{\mathrm{d}x}{\mathrm{d}t}=v_0$$

The resulting particular solution would be given by displacement function denoted by $x=h(t)$. The domain of h would be the time interval of interests in the problem. In summary, if we know the initial displacement and velocity of the particle, and also the DE governing the motion, a solution of the problem consists of the displacement function, which gives the position of the partile at variable time t. The DE and the initial conditions furnish a mathematical model for the physical situation.

【Example 4. 6. 5】 A particle moves on the x axis with acceleration $a=6t-4$ft/sec^2. Find the position and velocity of the particle at $t=3$ if the particle is at the origin and had velocity 10ft/sec when $t=0$.

Solution Ⅰ. We first solve the initial-value problem:

$$a=\frac{\mathrm{d}v}{\mathrm{d}t}=\frac{\mathrm{d}^2x}{\mathrm{d}t^2}=6t-4;\quad v=10, x=0, t=0$$

$$v=3t^2-4t+c$$

$$10=0-0+c$$

and

$$\frac{\mathrm{d}x}{\mathrm{d}t}=v=3t^2-4t+10 \tag{4. 6. 4}$$

$$x=t^3-2t^2+10t+k$$

$$0=0-0+k$$

$$x=t^3-2t^2+10t \tag{4. 6. 5}$$

Substituting $t=3$ in equation (4. 6. 4) and equation (4. 6. 5), we obtain $v_3=25$ft/sec and $x_3=39$ft.

Solution Ⅱ.

$$\int_{10}^{v}\mathrm{d}z=\int_0^t(6w-4)\mathrm{d}w$$

$$v-10=3t^2-4t$$

$$v_3=27-12+10=25\text{ft/sec}$$

and

$$\frac{\mathrm{d}x}{\mathrm{d}t}=3t^2-4t+10$$

$$\int_0^{3x^2}\mathrm{d}u=\int_0^3(3t^2-4t+10)\mathrm{d}t$$

$$x_3=t^3-2t^2+10t\,|_0^3=39\text{ft}$$

4.7 Exercises

1. Find a general solution of each of the following DE

(1) $y'-e^x=0$ (2) $y''-e^x=0$ (3) $y''-\cos x=0$

(4) $y'''-x+1=0$ (5) $y'-x\ln x=0$

2. Give the order of each of the following DE and state whether it is linear or nonlinear

(1) $\frac{d^2y}{dx^2}+xy=0$ (2) $\frac{dy}{dt}+t^2y=0$ (3) $\frac{d^3x}{dy^3}+\cos y=0$

(4) $\frac{d^3y}{dx^3}+\cos y=0$ (5) $\frac{d^2r}{d\theta^2}+r\frac{dr}{d\theta}=0$ (6) $x^2y+xy'+x^2-1=0$

(7) $\frac{dy}{dx}=x+y^2$ (8) $\frac{dy}{dx}=x^2+y$

3. Find a DE having the function defined by $y=e^t+e^{-t}$ as a solution.

4. Find the unique solution of $y'^2+4y^2=0$

5. Find the unique solution of $y'^2+3|y|=0$ on $(-\infty,+\infty)$

6. Find a complete solution of

(1) $y'=6x^2-2x$ on $(-\infty,+\infty)$

(2) $y'=4\tan x\sec^2 x$ on $\left(-\frac{\pi}{2}, \frac{\pi}{2}\right)$

(3) $y'=\ln x$ on $(0, +\infty)$

7. Solve $y'=|x|$ on $(-\infty,+\infty)$ given that (1) $y=6$ when $x=2$ and (2) $y=6$ when $x=-2$.

8. Find the general solution of the given differential equation.

(1) $\frac{dy}{dt}+y\cos t=0$ (2) $\frac{dy}{dt}+\frac{2t}{1+t^2}y=\frac{1}{1+t^2}$ (3) $\frac{dy}{dt}+t^2y=1$

9. Find the solution of the given initial-value problem.

(1) $\frac{dy}{dt}+\sqrt{1+t^2}\,e^{-t}y=0$, $y(0)=1$ (2) $\frac{dy}{dt}-2ty=t$, $y(0)=1$

(3) $\frac{dy}{dt}+y=\frac{1}{1+t^2}$, $y(1)=2$

10. Show that the following DE are not exact.

(1) $x^2y\,dx+x^3y^2\,dy=0$

(2) $x^2\,dx+y^2x\,dx=0$

(3) $-y\,dx+x\,dy=0$

(4) $(x^2+2xy)dx+(y^2-2xy)dy=0$

(5) $(4x\sin y)dx+(x^2\cos y)dy=0$

(6) $(2x^2y+y^2e^x)dx+(x^2+2ye^x)dy=0$

11. Show that the following DE are exact.

(1) $y^2dx+2xydy=0$

(2) $\ln y dx+xy^{-1}dy=0$

(3) $(3x^2\sin y)dx+(x^3\cos y)dy=0$

(4) $(ye^x-\sin y)dx+(e^x-x\cos y)dy=0$

12. Show that the following DE are exact. Solve by inspection.

(1) $y^2dx+2xydy=0$

(2) $e^y dx+xe^y dy=0$

13. Solve

(1) $dx-2xydy=0$

(2) $xdx-3dy=0$

(3) $\dfrac{dx}{y(1+x^2)}-2dy=0$

(4) $\dfrac{dy}{dx}=e^{y-x}$

(5) $2x(1+y^2)dx+dy=0$

(6) $\sqrt{1-y^2}dx-x\sqrt{x^2-1}dy=0$

14. Solve the initial-value problem:

(1) $(2xy+2x)dx-dy=0$, $y(0)=0$

(2) $dx=4\sqrt[3]{y(x-1)}dy$; $y(1)=0$

(3) $\dfrac{dy}{dx}=3y$; $y(0)=1$

(4) $\dfrac{dy}{dx}=\dfrac{3y-xy}{x}$, $y(1)=3e^{-1}$

15. Solve:

(1) $y/xdx+dy=0$

(2) $ydx-xdy=0$

(3) $2x(x^2+y^2)dx-xdy+ydx=0$

(4) $xdx+(y+e^y\sqrt{x^2+y^2})dy=0$

16. Find $(M_y-N_x)/N$ and solve:

(1) $(2y+4)dx+xdy=0$

(2) $(3y+8x)dx+xdy=0$

(3) $(x^2+2x+y)dx+dy=0$

(4) $(y+2x\cos^2 x\sin y)dx+(\sin x\cos x+x^2\cos^2 x\cos y)dy=0$

17. Solve the following initial-value problems.

(1) $y'=6x$; $x=0$, $y=4$

(2) $y'=-2x$; $x=0$, $y=7$

(3) $y'=4$; $x=0$, $y=-3$

(4) $y'=x-3$; $x=0$, $y=8$

(5) $y=x^2-2x-3$; $x=3$, $y=1$

(6) $y'=3x^2+4$; $x=-2$, $y=3$

18. Solve the following initial-value problem

$xy'-2=0$; $x=1$, $y=2\ln 3$

Find (1) y when $x=2$, and (2) x when $y=0$.

19. Solve the following two-point boundary-value problems.

(1) $y''=2$; $x=0$, $y=-1$; $x=1$, $y=5$

(2) $y''=12x$; $x=0$, $y=-3$; $x=1$, $y=0$

(3) $y''=-6x+2$; $x=1$, $y=-3$; $x=3$, $y=-21$

(4) $y''=12x^2$; $x=-1$, $y=-3$; $x=2$, $y=21$

Chapter 5

Second and higher order differential equations

5.1 Algebraic properties of solutions

A second-order differential equation is an equation of the form

$$\frac{d^2y}{dt^2}=f\left(t,y,\frac{dy}{dt}\right). \tag{5.1.1}$$

For example, the equation

$$\frac{d^2y}{dt^2}=\sin t+3y+\left(\frac{dy}{dt}\right)^2$$

is a second-order differential equation. A function $y=y(t)$ is a solution of equation (5.1.1) if $y(t)$ satisfies the differential equation; that is

$$\frac{d^2y(t)}{dt^2}=f\left(t,y(t),\frac{dy(t)}{dt}\right)$$

Thus, the function $y(t)=\cos t$ is a solution of the second-order equation $d^2y/dt^2=-y$ since $d^2y(\cos t)/dt^2=-\cos t$. Second-order differential equations arise quite often in applications. The most famous seconder-order differential equation is Newton's second law of motion

$$m\frac{d^2y}{dt^2}=F\left(t,y,\frac{dy}{dt}\right)$$

which governs the motion of a particle of mass m moving under the influence of a force F. In this equation, m is the mass of the particle, $y=y(t)$ is its position at time t, dy/dt is its velocity, and F is the total force acting on the particle. As the notation suggests, the force F may depend on the position and velocity of the particle, as well as on time. In addition of the equation (5.1.1), we will often impose initial conditions on $y(t)$ of the form

$$y(t_0)=y_0, \quad y'(t_0)=y'_0 \tag{5.1.2}$$

The differential equation (5.1.1) together with the initial conditions (5.1.2) is referred to as an initial-value problem. For example, let $y(t^*)$ denote the position at time t of a particle moving under the influence of gravity. Then $y(t)$ satisfies the initial-value problem

$$\frac{d^2y}{dt^2}=-g; \quad y(t_0)=y_0, \quad y'(t_0)=y'_0$$

where y_0 is the initial position of the particle and y' is the initial velocity of the particle. Second-order differential equations are extremely difficult to solve. This should not com as a great surprise to us after our experience with first-order equations. we will only succeed in solving the special differential equation

$$\frac{d^2y}{dt^2}+p(t)\frac{dy}{dt}+q(t)=g(t) \tag{5.1.3}$$

Fortunately, though, many of the second-order equations that arise in applications are of this form.

The differential equation (5.1.3) is called a seconder-order linear differential equation. We single out this equation and call it linear because both y and dy/dt appear by themselves. For example, the differential equations

$$\frac{d^2y}{dt^2}+3t\frac{dy}{dt}+(\sin t)y=e^t$$

and

$$\frac{d^2y}{dt^2}+e^t\frac{dy}{dt}+2y=1$$

are linear, while the differential equations

$$\frac{d^2y}{dt^2}+3t\frac{dy}{dt}+\sin y=t^3$$

and

$$\frac{d^2y}{dt^2}+\left(\frac{dy}{dt}\right)^2=1$$

are both nonlinear, due to the presence of the $\sin y$ and $(dy/dt)^2$ terms, respectively.

We consider first the second-order linear homogeneous equation

$$\frac{d^2y}{dt^2}+p(t)\frac{dy}{dt}+q(t)y=0 \tag{5.1.4}$$

which is obtained from equation (5.1.3) by setting $g(t)=0$. It is certainly not obvious at this point how to find all the solutions of equation (5.1.3), or how to solve initial-value problem

$$\frac{d^2y}{dt^2}+p(t)\frac{dy}{dt}+q(t)y=0';\ y(t_0)=y_0,\ y'(t_0)=y'(0) \qquad (5.1.5)$$

Therefore, before trying to develop any elaborate procedures for solving equation (5.1.4), we should first determine whether it actually has a solution. This information is contained in the following theorem.

Theorem 5.1.1. **(Existence-uniqueness Theorem).** Let the function $p(t)$ and $q(t)$ be continuous in the open interval $\alpha<t<\beta$. Then, there exists one, and only one function $y(t)$ satisfying the differential equation (5.1.3) on the entire interval $\alpha<1<\beta$ and the prescribed initial conditions $y(t_0)=y_0$, $y'(t_0)=y'(t_0)$. In particular, any solution $y=y(t)$ of (5.1.3) which satisfies $y(t_0)=0$ and $y'(t_0)=0$ at some time $t=t_0$ must be identically zero. Theorem 5.1.1 is an extremely important theorem for us. On the one hand, it is our hunting license to find the unique solution $y(t)$ of equation (5.1.4). And, on the other hand, we will actually use Theorem 5.1.1 to help us find all the solutions of equation (5.1.3).

We begin our analysis of equation (5.1.3) with the important observation that the left-hand side $y''+p(t)y'+q(t)y\left(y'=\frac{dy}{dt}, y''=\frac{d^2y}{dt^2}\right)$ of the differential equation can be viewed as defining a "function of a function": with each function y having two derivatives, we associate another function, which we will call $L[y]$, by the relation

$$L[y](t)=y''(t)+p(t)y'(t)+q(t)y(t)$$

In mathematical terminology, L is an operator which operates on functions; that is, there is a prescribed recipe for associating with each function y a new function $L[y]$.

【Example 5.1.1】 Let $p(t)=0$ and $q(t)=t$. Then

$$L[y](t)=y''(t)+ty(t)$$

If $y(t)=\cos t$, then $L[y](t)=(\cos t)''+t\cos t=(t-1)\cos t$,
and if $y(t)=t^3$, then $L[y](t)=(t^3)''+t(t^3)=t^4+6t$.

Thus, the operator L assigns the function $(t-1)\cos t$ to the function $\cos t$, and the function $6t+t^4$ to the function t^3.

The concept of an operator acting on functions, or a "function of a function" is analogous to that f a function of a single variable t. Recall the definition of a function f on an interval I: with each number t in I we associate a new number called $f(t)$. In an exactly analogous manner, we associate with each function y having two derivatives a new function called $L[y](t)$. This is an extremely sophisticates mathematical concept, because in a certain sense, we are treating a function exactly as we do a point. Admittedly, this is unite difficult to grasp.

It's not surprising, therefore, that the concept of a "function of a function" was not developed till the beginning of this century, and that many of the "high powered" theorems of mathematical analysis were proved only after this concept was mastered.

We now derive several important properties of the operator L. which we will use to great advantage shortly.

Property 5.1.1. $L[cy]=cL[y]$, for any constant c.

Proof.

$$L[cy](t)=(cy)''(t)+p(t)(cy)'(t)+q(t)(cy)(t)$$
$$=cy''(t)+cp(t)y'(t)+cq(t)y(t)=c[y''(t)]$$

The meaning of Property 5.1.1 is that the operator L assigns to the function (cy) equal to c times the function it assigns to y. For example, let

$$L[y](t)=y''(t)+6y'(t)-2y(t)$$

This operator L assigns the function

$$(t^2)''+6(t^2)-2(t^2)=2+12t-2t^2$$

to the function t^2. Hence, L must assign the function $5(2+12t-2t^2)$ to the function $5t^2$.

Property 5.1.2. $L[y_1+y_2]=L[y_1]+L[y_2]$.

Proof.

$$L[y_1+y_2](t)=(y_1+y_2)''(t)+p(t)(y_1+y_2)'(t)+q(t)(y_1+y_2)(t)$$
$$=y_1''(t)+y_2''(t)+p(t)y_1'(t)+p(t)y_2'(t)+q(t)y_1(t)+q(t)y_2(t)$$
$$=[y_1''(t)+p(t)y_1'(t)+q(t)y_1(t)]+[y_2''(t)+p(t)y_2'(t)+q(t)y_2(t)]$$
$$=L[y_1](t)+L[y_2](t)$$

The meaning of Property 5.1.2 is that the operator L assigns to the function y_1+y_2 equal to the sum of the functions it assigns to y_1 and y_2. For example, let

$$Ly[t](t)=y''(t)+2y'(t)-y(t)$$

This operator L assigns the function

$$(\cos t)''+2(\cos t)'-\cos t=-2\cos t-2\sin t$$

to the function $\cos t$, and the function

$$(\sin t)''+2(\sin t)'-\sin t=2\cos t-2\sin t$$

to the function $\sin t$. Hence, L assigns the function

$$(-2\cos t-2\sin t)+2\cos t-2\sin t=-4\sin t$$

to the function $\sin t+\cos t$.

Definition 5.1.1. An operator L which assigns functions to functions and which satisfies Properties 5. 1. 1 and 5. 1. 2 is called a linear operator[1]. All other operators are nonlinear. An example of a nonlinear operators

$$L[y](t)=y''(t)-2t[y(t)]^4.$$

This operator assigns the function

$$\left(\frac{1}{t}\right)''-2t\left(\frac{1}{4}\right)^4=\frac{2}{t^3}-\frac{2}{t^3}=0$$

to the function $1/t$, and the function

$$\left(\frac{c}{t}\right)''-2t\left(\frac{c}{t}\right)^4=\frac{2c}{t^3}-\frac{2c^4}{t^3}=\frac{2c(1-c^3)}{t^3}$$

to the function c/t. Hence, for $c=0$, 1 and $y(t)=1/t$, we see that

$$L[cy]=cL[y]$$

The usefulness of Properties 5. 1. 1 and 5. 1. 2 lies in the observation that the solutions $y(t)$ of the differential equation (5. 1. 2) are exactly those functions y for which

$$L[y](t)=y''(t)+p(t)y'(t)+q(t)y(t)=0$$

In other words, the solutions $y(t)$ of equation (5. 1. 4) are exactly those functions y to which the operator L assigns the zero function. Hence, if $y(t)$ is a solution of equation (5. 1. 4) then so is $cy(t)$, since

$$L[cy](t)=cL[y](t)=0$$

If $y_1(t)$ and $y_2(t)$ are solutions of equation (5. 1. 4), then $y_1(t)+y_2(t)$ is also a solution of equation (5. 1. 4), since

$$L[y_1+y_2](t)=L[y_1](t)+L[y_2](t)=0+0=0$$

Combining Properties 5. 1. 1 and 5. 1. 2, we see that all linear combinations

$$c_1y_1(t)+c_2y_2(t)$$

of solutions of equation (5. 1. 4) are again solutions of equation (5. 1. 4).

The preceding argument shows that we can use our knowledge of two solutions $y_1(t)$ and $y_2(t)$ of equation (5. 1. 4) to generate infinitely many other solutions. This statement has some very interesting implications.

Consider, for example, the differential equation

$$\frac{d^2y}{dt^2}+y=0 \tag{5. 1. 6}$$

Two solutions of equation (5. 1. 6) are $y_1(t)=\cos t$ and $y_2(t)=\sin t$. Hence,

$$y(t)=c_1\cos t+c_2\sin t \tag{5. 1. 7}$$

is also a solution of equation (5. 1. 6), for every choice of constants c_1 and c_2. Now, Equation (5. 1. 7) contains two arbitrary constants. It is natural to suspect, there-

---

[1] linear operator 线性算子

fore, that this expression represents the general solution of equation (5.1.6); that is, every solution $y(t)$ of equation (5.1.6) must be of the form equation (5.1.7).

That is indeed the case, as we now show. Let $y(t)$ be any solution of equation (5.1.6).

By the existence-uniqueness theorem, $y(t)$ exists for all t.

$$\text{Let } y(0)=y_0, y'(0)=y_0'$$

and consider the function

$$\phi(t)=y_0\cos t+y_0'\sin t$$

This function is a solution of equation (5.1.6) since it is a linear combination of solutions of equation (5.1.6). Moreover, $\phi(0)=y_0$ and $\phi'(0)=y_0'$. Thus, $y(t)$ and $\phi(t)$ satisfy the same second-order linear homogeneous equation and the same initial conditions. Therefore, by the uniqueness part of Theorem 1, $y(t)$ must be identically equal to $\phi(t)$, so that

$$y(t)=y_0\cos t+y_0'\sin t$$

Thus, equation (5.1.7) is indeed the general solution of equation (5.1.6).

Let us return now to the general liner equation (5.1.4). Suppose, in some manner, that we manage to find two solutions y_1 and y_2 of equation (5.1.4). Then every function

$$y(t)=c_1y_1(t)+c_2y_2(t) \tag{5.1.8}$$

is again a solution of equation (5.1.4). Does the expression (5.1.8) represent the general solution of equation (5.1.4)? That is to say, does every solution $y(t)$ of (5.1.4) have the form (5.1.8)? The following theorem answers this question.

Theorem 5.1.2. Let y_1 and y_2 be two solutions of equation (5.1.4) on the interval $\alpha<t<\beta$, with $y_1(t)y_2'(t)-y_1'(t)y_2(t)$ unequal to zero in this interval. Then, $y(t)=c_1y_1(t)+c_2y_2(t)$ is the general solution of equation (5.1.4).

Proof. Let $y(t)$ be any solution of equation (5.1.4). We must find constants c_1 and c_2 such that $y(t)=c_1y_1(t)+c_2y_2(t)$. To this end, pick a time t_0 in the interval (α,β) and let y_0 and y_0' denote the values of y and y' at $t=t_0$. The constants c_1 and c_2 if they exists, must satisfy the two equations

$$c_1y_1(t_0)+c_2y_2(t_0)=y_0$$
$$c_1y_1'(t_0)+c_2y_2'(t_0)=y_0'$$

Multiplying the first equation by $y_2'(t_0)$, the second equation by $y_2(t_0)$ and subtracting gives

$$c_1[y_1(t_0)y_2'(t_0)-y_1'(t_0)y_2(t_0)]=y_0y_2'(t_0)-y_0'y_2(t_0)$$

Similarly, multiplying the first equation by $y_1'(t_0)$, the second equation by $y_1(t_0)$ and subtracting gives

$$c_2[y_1'(t_0)y_2(t_0)-y_1(t_0)y_2'(t_0)]=y_0y_1'(t_0)-y_0'y_1(t_0)$$

Hence,

$$c_1=\frac{y_0y_2'(t_0)-y_0'y_2(t_0)}{y_1(t_0)y_2'(t_0)-y_1'(t_0)y_2(t_0)}$$

and

$$c_2=\frac{y_0y_1'(t_0)-y_0'y_1(t_0)}{y_1(t_0)y_2'(t_0)-y_1'(t_0)y_2(t_0)}$$

if

$$y_1(t_0)y_2'(t_0)-y_1'(t_0)y_2(t_0)\neq 0$$

Now, let

$$\phi(t)=c_1y_1(t)+c_2y_2(t)$$

for this choice of constant c_1, c_2. We know that $\phi(t)$ satisfies equation (5.1.4), since it is a linear combination of solutions of equation (5.1.4). Moreover, by construction, $\phi(t_0)=y_0$ and $\phi'(t_0)=y_0'$. Thus, $y(t)$ and $\phi(t)$ satisfy the same second-order linear homogeneous equation and the same initial conditions. Therefore, by the uniqueness part of Theorem 5.1.1, $y(t)$ must be identically equal to $\phi(t)$; that is,

$$y(t)=c_1y_1(t)+c_2y_2(t),\alpha<t<\beta$$

Theorem 5.1.2. is an extremely useful theorem since it reduces the problem of finding all solutions of equation (5.1.4), of which there are infinitely many, to the much simple problem of finding just two solutions $y_1(t)$, $y_2(t)$. The only condition imposed on the solutions $y_1(t)$ and $y_2(t)$ is that the quantity $y_1(t)y_2'(t)-y_1'(t)y_2(t)$ be unequal to zero for $\alpha<t<\beta$. When this is the case, we say that $y_1(t)$ and $y_2(t)$ are a fundamental set of solutions of equation (5.1.4), since all other solutions of equation (5.1.4) can be obtained by taking linear combinations of $y_1(t)$ and $y_2(t)$.

Definition 5.1.2. The quantity $y_1(t)y_2'(t)-y_1'(t)y_2(t)$ is called the Wronskian[1] of $y_1(t)$ and $y_2(t)$, and is denoted by $W(t)=W[y_1,y_2](t)$.

Theorem 5.1.2. requires that $W[y_1,y_2](t)$ be unequal to zero at all points in the interval (α,β). In actual fact, the Wronskian of any two solutions $y_1(t)y_2(t)$ of equation (5.1.4) is either identically zero, or is never zero, as we now show.

Theorem 5.1.3. Let $p(t)$ and $q(t)$ be continuous in the interval $\alpha<t<\beta$ and let $y_1(t)$ and $y_2(t)$ be two solutions of equation (5.1.4). Then, $W[y_1,y_2](t)$ is either identically zero, or is never zero, on the interval $\alpha<t<\beta$.

We prove Theorem 5.1.3 with the aid of the following lemma.

---

[1] Wronskian 朗斯基行列式

Lemma 5.1.1. Let $y_1(t)$ and $y_2(t)$ be two solutions of the linear differential equation $y''+p(t)y'(t)+q(t)y=0$.

Then, their Wronskian

$$W(t)=W[y_1,y_2](t)=y_1(t)y_2'(t)-y_1'(t)y_2(t)$$

satisfies the first-order differential equation

$$W'+p(t)W=0$$

Proof. Observe that

$$W'(t)=\frac{\mathrm{d}}{\mathrm{d}t}(y_1y_2'-y_1'y_2)=y_1y_2''+y_1'y_2'-y_1'y_2'-y_1''y_2=y_1y_2''-y_1''y_2$$

Since y_1 and y_2 solutions of $y''+p(t)y'+q(t)y=0$, we know that

$$y_2''=-p(t)y_2'-q(t)y_2$$

and

$$y_1''=-p(t)y_1'-q(t)y_1$$

Hence,

$$\begin{aligned}W'(t)&=y_1[-p(t)y_2'-q(t)y_2]-y_2[-p(t)y_1'-q(t)y_1]\\&=-p(t)[y_1y_2'-y_1'y_2]=-p(t)W(t)\end{aligned}$$

We can now give a very simple proof of Theorem 5.1.3.

Proof. Pick any t_0 in the interval (α,β). From Lemma 5.1.1,

$$W[y_1,y_2](t)=W[y_1,y_2](t_0)\exp\left(-\int_{t_0}^{t}p(s)\mathrm{d}s\right)$$

Now, $\exp\left(-\int_{t_0}^{t}p(s)\mathrm{d}s\right)$ is equal to zero for $\alpha<t<\beta$. Therefore, $W[y_1,y_2](t)$ is either identically zero, or is never zero.

The simplest situation where the Wronskian of functions $y_1(t)$, $y_2(t)$ vanishes identically is when one of the functions is identically zero. More generally, the Wronskian of two functions y_1, y_2 vanishes identically if one of the functions is a constant multiple of the other. If $y_2=cy_1$, then

$$W[y_1,y_2](t)=y_1(cy_1)'-y_1(cy_1)=0.$$

Conversely, suppose that the Wronskian of two solutions $y_1(t)$, $y_2(t)$ of equation (5.1.4) vanishes identically. Then t_0, one of these solutions must be a constant multiple of the others, as we now show.

Theorem 5.1.4. Let $y_1(t)$ and $y_2(t)$ be two solutions of equation (5.1.4) on the interval $\alpha<t<\beta$, and suppose that $W[y_1,y_2](t_0)=0$ for some in this interval. Then, one of these solutions is a constant multiple of the other.

Proof Ⅰ. Suppose that $W[y_1,y_2](t_0)=0$. Then, the equations

$$c_1y_1(t_0)+c_2y_2(t_0)=0$$
$$c_1y_1'(t_0)+c_2y_2'(t_0)=0$$

have a nontrivial solution c_1, c_2; that is, a solution c_1, c_2 with $|c_1+|c_2|\neq 0$ Let $y(t)=c_1y_1(t)+c_2y_2(t)$, for this choice of constants c_1, c_2. We know that $y(t)$ is a solution of equation (5.1.4), since it is a linear combination of $y_1(t)$ and $y_2(t)$. Moreover, by construction, $y(t)$ is identically zero, so that

$$c_1y_1(t)+c_2y_2(t)=0, \alpha<t<\beta$$

If $c_1\neq 0$, then $y_1(t)=-(c_2/c_1)y_2(t)$, and if $c_2\neq 0$, then $y_2(t)=-(c_1/c_2)y_1(t)$. In either case, one of these solutions is a constant multiple of the other.

Proof Ⅱ. Suppose that $W[y_1,y_2](t_0)=0$. Then, by Theorem 5.1.3, $W[y_1,y_2](t)$ is identically zero. Assume that

$$y_1(t)y_2(t)\neq 0 \text{ for } \alpha<t<\beta$$

Then, dividing both sides of the equation

$$y_1(t)y_2'(t)-y_1'(t)y_2(t)=0$$

by $y_1(t)y_2(t)$ gives

$$\frac{y_2'(t)}{y_2(t)}-\frac{y_1'(t)}{y_1(t)}=0$$

This equation implies that $y_1(t)=cy_2(t)$ for some constant c.

Next, suppose that $y_1(t)y_2(t)$ is zero at some point $t=t^*$ in the interval $\alpha<t<\beta$. Without loss of generality, we may assume that $y_1(t^*)=0$, since otherwise we can relabel $y_1(t)$ and $y_2(t)$. In this case it is simple to show that either $y_1(t)\equiv 0$, or $y_2(t)=[y_2'(t^*)/y_1'(t^*)]y_1(t)$. This completes the proof of Theorem 5.1.4.

Definition 5.1.3. The functions $y_1(t)$ and $y_2(t)$ are said to be linearly dependent[1] on an interval I if one of these functions is a constant of the other on I. The functions $y_1(t)$ and $y_2(t)$ are said to be linearly independent[2] on an interval I if they are not linearly on I.

Corollary to Theorem 5.1.4. Two solutions $y_1(t)$ and $y_2(t)$ of equation (5.1.4) are linearly independent on the interval $\alpha<t<\beta$, if and only if, their Wronskian is unequal to zero on this interval. Thus, two solutions $y_1(t)$ and $y_2(t)$ form a fundamental set of solutions of equation (5.1.4) on the interval $\alpha<t<\beta$, and only if, they are linearly independent on this interval.

5.2 Linear equations with constant coefficients

We consider now the homogeneous linear second-order equation with constant coeffi-

---

❶ linearly dependent 线性相关

❷ linearly independent 线性无关

cients

$$L[y]=a\frac{d^2y}{dt^2}+b\frac{dy}{dt}+cy=0 \tag{5.2.1}$$

where a, b and c are constants, with $a\neq 0$. Theorem 5.1.2 of Section 5.1 tells us that we need only two independent solutions y_1 and y_2 of equation (5.2.1); all other solutions of equation (5.2.1) are then obtained by taking linear combinations of y_1 and y_2. Unfortunately, Theorem 5.1.2 doesn't tell us how to find two solutions of equation (5.2.1). Therefore, we will try an educated guess. To this end, observe that a function $y(t)$ is a solution of equation (5.2.1) if a constant times its second derivative, plus a third constant times itself is identically zero. In other words, the three terms ay', by' and cy must cancel each other. In general, this can only occur if the three functions $y(t)$, $y'(t)$, and $y''(t)$ are of the same type. For example, the functions $y(t)=t^5$ can never be a solution of equaton (5.2.1) since the three terms $20at^3$, $5bt^4$ and ct^5 are polynomials in t of different degree, and therefore cannot cancel each other. On the other hand, the function $y(t)=e^{rt}$, r constant, has the property that both $y'(t)$ and $y''(t)$ are multiples of $y(t)$. This suggests that we try $y(t)=e^{rt}$ as a solution of equation (5.2.1). Computing

$$L[e^{rt}]=a(e^{rt})''+b(e^{rt})'+c(e^{rt})=(ar^2+br+c)e^{rt}$$

We see that $y(t)=e^{rt}$ is solution of (5.2.1). If, and only if

$$ar^2+br+c=0 \tag{5.2.2}$$

Equation (5.2.2) is called the characteristic equation of (5.2.1). It has two roots r_1, r_2 given by the quadratic formula

$$r_1=\frac{-b+\sqrt{b^2-4ac}}{2a}, r_2=\frac{-b-\sqrt{b^2-4ac}}{2a}$$

If b^2-4ac is positive, then r_1 and r_2 are real and distinct. In this case, $y_1(t)=e^{r_1t}$ and $y_2(t)=e^{r_2t}$ are two distinct solutions of equation (5.2.1). These solutions are clearly linearly independent (on any interval I), since e^{r_2t} is obviously not a constant multiple of e^{r_1t} for $r_2\neq r_1$.

It is necessary to consider three cases; r_1 and r_2 may be real and unequal, r_1 and r_2 may be real and equal, and r_1 and r_2 may be complex numbers of the form $\alpha\pm i\beta(\beta\neq 0)$.

Case Ⅰ: Roots real and unequal

Let us assume that r_1 and r_2 are distinct real roots of equation (5.2.2). It is easy to show that $y_1=e^{r_1x}$ and $y_2=e^{r_2x}$ define solutions of equation (5.2.1). For example,

$$(D^2+aD+b)e^{r_1x}=e^{r_1x}(r_1^2+ar_1+b)=0$$

Since r_1 is a root of the characteristic equation.

The solution y_1 and y_2 are linearly independent, since

$$\begin{aligned}W(y_1, y_2)&=e^{r_1x}(r_2e^{r_2x})-e^{r_2x}(r_1e^{r_1x})\\&=e^{(r_1+r_2)x}(r_2-r_1)\end{aligned}$$

cannot be identically zero on interval I due to the fact that $e^{(r_1+r_2)x}$ and r_2-r_1 are both different from zero.

Hence, $y=c_1e^{r_1x}+c_2e^{r_2x}$ defines a complete solution of (5.2.1), valid on $(-\infty,+\infty)$.

【Example 5.2.1】 Find a complete solution of $(D^2+D-6)y(x)=0$, $r_1=-3$.

Solution. The characteristic equation $r^2+r-6=(r+3)(r-2)=0$ has roots and $r_2=2$ and hence the DE has a complete solution given by

$$y=c_1e^{-3x}+c_2e^{2x}$$

【Example 5.2.2】 Find a complete solution of $(D^2+5D)y(x)=0$.

Solution. The characteristic equation $r^2+5r=0$ has roots $r_1=0$ and $r_2=-5$ and hence the DE has a complete solution given by

$$y=c_1e^{0x}+c_2e^{-5x} \text{ or } y=c_1+c_2e^{-5x}$$

【Example 5.2.3】 Find a complete solution of $(D^2-4D+1)y(x)=0$ for which $y(0)=0, y'(0)=\sqrt{3}$.

Solution. The characteristic equation $r^2-4r+1=0$ has roots $r_1=2+\sqrt{3}$ and $r_2=2-\sqrt{3}$ and hence the DE has a complete solution given by

$$y=c_1\exp(2+\sqrt{3})x+c_2\exp(2-\sqrt{3})x$$

$$y'=c_1(2+\sqrt{3})\exp(2+\sqrt{3})x+c_2(2-\sqrt{3})\exp(2-\sqrt{3})x$$

We obtain

$$c_1+c_2=0$$

$$c_1(2+\sqrt{3})+c_2(2-\sqrt{3})=\sqrt{3}$$

yielding $c_1=\frac{1}{2}$, $c_2=-\frac{1}{2}$.

The required solution is given by

$$y=\frac{1}{2}[\exp(2+\sqrt{3})x-\exp(2-\sqrt{3})x]$$

【Example 5.2.4】 Find the general solution of the equation

$$\frac{d^2y}{dt^2}+5\frac{dy}{dt}+4y=0 \qquad (5.2.3)$$

Solution. The characteristic equation $r^2+5r+4=(r+4)(r+1)=0$ has two distinct roots $r_1=-4$ and $r_2=-1$. Thus, $y_1=e^{-4t}$ and $y_2(t)=e^{-t}$ form a fundamental set of solutions of equation (5.2.3), and every solution $y(t)$ of equation (5.2.3) is of the form

$$y(t)=c_1 e^{-4t}+c_2 e^{-t}$$

for some choice of constant c_1, c_2.

【Example 5.2.5】 Find the solution $y(t)$ of the initial-value problem

$$\frac{d^2 y}{dt^2}+4\frac{dy}{dt}-2y=0; y(0)=1, y'(0)=2 \tag{5.2.4}$$

Solution. The characteristic equation $r^2+4r-2=0$ has 2 roots

$$r_1=\frac{-4+\sqrt{16+8}}{2}=-2+\sqrt{6}$$

and

$$r_2=\frac{-4-\sqrt{16+8}}{2}=-2-\sqrt{6}$$

Hence, $y_1(t)=e^{r_1 t}$ and $y_2(t)=e^{r_2 t}$ are a fundamental set of solutions of $y''+4y'-2y=0$ so that

$$y(t)=c_1 e^{(-2+\sqrt{6})t}+c_2 e^{(-2-\sqrt{6})t}$$

for some choice of constants c_1, c_2 The constants c_1, c_2 are determined from the initial conditions

$$c_1+c_2=1 \text{and} (-2+\sqrt{6})c_1+(-2-\sqrt{6})c_2=2$$

From the first equation, $c_2=1-c_1$. Substituting this value of c_2 into the second equation gives

$$(-2+\sqrt{6})c_1-(2+\sqrt{6})(1-c_1)=2, \text{or } 2\sqrt{6}c_1=4+\sqrt{6}$$

Therefore, $c_1=2/\sqrt{6}+\frac{1}{2}$, $c_2=1-c_1=\frac{1}{2}-2\sqrt{6}$, and

$$y(t)=\left(\frac{1}{2}+\frac{2}{\sqrt{6}}\right)e^{-2+\sqrt{6}t}+\left(\frac{1}{2}-\frac{2}{\sqrt{6}}\right)e^{-(2+\sqrt{6})t}$$

Case Ⅱ: Roots real and equal

When the discriminant a^2-4b of the characteristic equation is zero, the roots r_1 and r_2 of the characteristic equation are real and equal. In this case, $y_1=e^{r_1 x}$ defines a solution of (5.2.1) as in Case Ⅰ, but we lack a second solution y_2 such that y_1 and y_2 are linearly independent. We note, however, that the DE $D^2 y(x)=0$ is of this type, since it has characteristic equation $r^2=0$, with roots $r_1=r_2=0$. But $D^2 y(x)=0$ is easily solved by integrating twice with respect to obtain the solution given by

$$y=c_1(1)+c_2x$$

This is a complete solution of $D^2y(x)=0$ since the functions defined by $y_1=1$ and $y_2=x$ are linearly independent.

Since $y_2=xy_1$ in this particular example, it is natural to test to see whether $y_2=xe^{r_1x}$ defines a solution of equation (5. 2. 1) when $r_1=r_2$.

The function y_2 is indeed a solution since

$$y_2'=r_1e^{r_1x}+e^{r_1x}, a=-(r_1+r_1)=-2r, b=r_1r_1=r_1^2$$

and

$$\begin{aligned}y_2''+ay_2'+by_2&=(r_1^2xe^{r_1x}+2r_1xe^{r_1x})+(-2r_1)(r_1xe^{r_1x}+e^{r_1x})+r_1^2xe^{r_1x}\\&=e^{r_1x}(r_1^2x+2r_1-2r_1^2x-2r_1+r_1^2x)\\&=0\end{aligned}$$

The function y_1 and y_2 are linearly independent on an arbitrary interval I since their ratio $y_2/y_1=x$ is not constant on I. Hence when $r_1=r_2$, the DE (5. 2. 1) has a complete solution by $y=c_1e^{r_1x}+c_2xe^{r_1x}$.

The expression for y_2 may also be found from the formula

$$y_2=y_1\int\frac{\exp(-\int p\,dx)}{y_1^2}dx$$

With $y_1=e^{r_1x}$ and $p=a$, this formula yields

$$y_2=e^{r_1x}\int\frac{\exp(-ax)}{e^{2r_1x}}dx$$

Since $a=-(r_1+r_2)=-2r_1$,

$$\begin{aligned}y_2&=e^{r_1x}\int-e^{2r_1x}e^{-ax}\,dx=e^{r_1x}\int e^{ax}e^{-ax}\,dx\\&=e^{r_1x}\int dx=xe^{r_1x}\end{aligned}$$

【Example 5. 2. 6】 Find a complete solution of $(D^2-8D+16)y(x)=0$.

Solution. The characteristic equation $r^2-8r+16=0$ has $r=4$ as a repeated root. Hence the DE has a complete solution given by $y=c_1e^{4x}+c_2xe^{4x}$.

【Example 5. 2. 7】 Solve the initial-value problem

$$(D^2+4D+4)y(x)=0; y(0)=4, y'(0)=-5.$$

Solution. The characteristic equation $r^2+4r+4=(r+2)^2=0$ has $r=-2$ as a repeated root.

Setting $x=0$ in

$$y=c_1e^{-2x}+c_2xe^{-2x}$$

and

$$y' = -2c_1 e^{-2x} - 2c_2 x e^{-2x} + c_2 e^{-2x}$$

we obtain
$$y(0) = c_1 = 4$$

and
$$y'(0) = -2(4) + c_2 = -5$$

From $c_1 = 4$ and $c_2 = 3$ we obtain the required solution

$$y = 4e^{-2x} + 3xe^{-2x} = e^{-2x}(4 + 3x)$$

Case Ⅲ: Roots not real

If the roots of the characteristic equation $r^2 + ar + b = 0$ are not real, then they must be complex conjugates, since a and b are real numbers. This situation prevails when the discriminant $a^2 - 4b$ of the characteristic equation is negative. Let these complex roots be denoted by $r_1 = \alpha + i\beta$ and $r_2 = \alpha - i\beta$, where α and β are real, $\beta \neq 0$, and $i = \sqrt{-1}$ is the complex imaginary unit having the property that $i^2 = -1$.

Since $a = -(r_1 + r_2) = -2\alpha$ and $b = r_1 r_2 = \alpha^2 - (i\beta)^2 = \alpha^2 + \beta^2$ the characteristic equation can be written in the form

$$r^2 - 2\alpha r + (\alpha^2 + \beta^2) = 0$$

If $\beta = 0$, then $r = \alpha$ is a double root of the characteristic equation $r^2 - 2\alpha r + \alpha^2 = 0$, and the DE reduces to $(D^2 - 2\alpha D + \alpha^2)y(x) = 0$. We know that this DE has a solution given by $y = e^{\alpha x}$.

If $\alpha = 0$, then is the characteristic equation, and the DE reduces to $(D^2 + \beta^2)y(x) = 0$. But it is seen by inspection that this DE has solution given by $y_1 = \sin\beta x$ and $y_2 = \cos\beta x$. These two special DE suggest that when α and β are both different from zero, we try f_1 and f_2 defines by $y_1 = f_1(x) = e^{\alpha x}\sin\beta x$ and $y_2 = f_2(x) = e^{\alpha x}\cos\beta x$ as solutions of equation (5.2.1).

The following computation, with $a = -2\alpha$ and $b = \alpha^2 + \beta^2$, verifies that f_1 is a solution of equation (5.2.1).

$$\begin{aligned}
y_1'' + ay_1' + by_1 &= y_1'' - 2\alpha y_1' + (\alpha^2 + \beta^2)y_1 \\
&= (-\beta^2 e^{\alpha x}\sin\beta x + \alpha\beta e^{\alpha x}\cos\beta x + \alpha\beta e^{\alpha x}\cos\beta x + \alpha^2 e^{\alpha x}\sin\beta x) \\
&\quad - 2\alpha(\beta e^{\alpha x}\cos\beta x + \alpha e^{\alpha x}\cos\beta x) + (\alpha^2 + \beta^2)e^{\alpha x}\sin\beta x \\
&= e^{\alpha x}(-\beta^2\sin\beta x + 2\alpha\beta\cos\beta x + \alpha^2\sin\beta x - 2\alpha\beta\cos\beta x - 2\alpha^2\sin\beta x \\
&\quad + \alpha^2\sin\beta x + \beta^2\sin\beta x) = 0
\end{aligned}$$

For a verification that f_2 is also a solution of equation (5.2.1).

The functions y_1 and y_2 are linearly independent on an arbitrary interval I since their ratio $y_2/y_1 = \cot\beta x$ is not constant on I.

Hence, when $\beta \neq 0$,

$$\begin{aligned}
y &= c_1 e^{\alpha x}\sin\beta x + c_2 e^{\alpha x}\cos\beta x \\
&= e^{\alpha x}(c_1\sin\beta x + c_2\cos\beta x) \qquad (5.2.5)
\end{aligned}$$

Defines a complete solution of equation (5.2.1).

Another way of obtaining equation (5. 2. 3) is to write

$$y_1 = e^{r_1 x} \text{ and } y_2 = e^{r_2 x}$$

as

$$y_1 = e^{(\alpha + i\beta)x} \text{ and } y_2 = e^{(\alpha - i\beta)x}$$

Then we proceed formally, without justifying our use of e^z when z is complex, and write

$$y_1 = e^{\alpha x} e^{i\beta x} \text{ and } y_2 = e^{\alpha x} e^{-i\beta x}$$

This step is suggested by the familiar law of exponents $e^{u+v} = e^u e^v$, known to hold when u and v are real numbers.

To assign a reasonable meaning to $e^{i\beta x}$ and $e^{-i\beta x}$, we recall that

$$e^z = \exp z = 1 + \frac{z}{1!} + \frac{z^2}{2!} + \cdots + \frac{z^{n-1}}{(n-1)!} + \cdots$$

For all real values of z. This is known as the Maclaurin series for e^z. We next set

$$e^{i\theta} = 1 + \frac{(i\theta)}{1!} + \frac{(i\theta)^2}{2!} + \cdots + \frac{(i\theta)^{n-1}}{(n-1)!} + \cdots$$

Replacement of i^2 by -1, $i^3 = i^2 i$ by $-i$, $i^4 = (i^2)^2$ by $+1$, $\cdots$, and $i^{4m+k} = i^{4m} i^k$ by i^k, where $k = 0, 1, 2$ or 3, we obtain

$$\begin{aligned} e^{i\theta} &= 1 + \frac{(i\theta)}{1!} - \frac{\theta^2}{2!} - \frac{i\theta^3}{3!} + \frac{\theta^4}{4!} \cdots \\ &= \left(1 - \frac{\theta^2}{2!} + \frac{\theta^4}{4!} \cdots\right) + i\left(\frac{(i\theta)}{1!} - \frac{i\theta^3}{3!} \cdots\right) \\ &= \cos\theta + i\sin\theta \end{aligned}$$

This is the famous Euler identity. Setting $\theta^1 = \beta x$ and $\theta^2 = -\beta x$, we obtain

$$e^{i\beta x} = \cos\beta x + i\sin\beta x$$

and

$$\begin{aligned} e^{-i\beta x} &= \cos(-\beta x) + i\sin(-\beta x) \\ &= \cos\beta x - i\sin\beta x \end{aligned}$$

These formulas yield

$$y_1 = e^{\alpha x}(\cos\beta x + i\sin\beta x) \text{ and } y_2 = e^{\alpha x}(\cos\beta x - i\sin\beta x)$$

The linear combinations

$$y_3 = \frac{1}{2i} y_1 - \frac{1}{2} y_2 \text{ and } y_4 = \frac{1}{2} y_1 + \frac{1}{2} y_2$$

yield

$$y_3 = e^{\alpha x} \sin\beta x \text{ and } y_4 = e^{\alpha x} \cos\beta x$$

and the linear combination $y = c_1 y_3 + c_4 y_4$ yields equation (5. 2. 3).

For a justification of the use of complex numbers in this development, see any standard reference on the theory of functions of a complex variable. From a practical point of view, the logical justification of a method by which a solution of a DE is obtained is not very important. If a solution obtained by any method whatsoever can be verified by substituting the solution into the DE, the solution has the same standing as one produced by a completely logical development.

【Example 5.2.8】 Find a complete solution of $(D^2-4D+13)y(x)=0$.

Solution. By the quadratic formula, the characteristic equation $r^2-4r+13=0$ has roots $r_1=2+3\mathrm{i}$ and $r_2=2-3\mathrm{i}$.

$$y=\mathrm{e}^{2x}(c_1\sin 3x+c_2\cos 3x)$$

【Example 5.2.9】 Find the solution of $(D^2+4)y(x)=0$ for which $y(0)=0$; $y'(0)=6$.

Solution. The characteristic equation $r^2+4=0$ has roots $r_1=0+2\mathrm{i}$ and $r_2=0-2\mathrm{i}$. A complete solution of $(D^2+4)y(x)=0$ is given by

$$y=\mathrm{e}^{0x}(c_1\sin 2x+c_2\cos 2x)$$

or

$$y=c_1\sin 2x+c_2\cos 2x$$

Substituting $x=0$ and $y=0$, we find that $c_1=3$. Hence the required particular solution is given by $y=3\sin 2x$.

5.3 The non-homogeneous equation

Now we turn our attention to the non-homogeneous equation

$$L[y]=\frac{\mathrm{d}^2y}{\mathrm{d}t^2}+p(t)\frac{\mathrm{d}y}{\mathrm{d}t}+q(t)y=g(t) \tag{5.3.1}$$

where the function $p(t)$, $q(t)$ and $g(t)$ are continuous on an open interval $\alpha<t<\beta$. An important clue as to the nature of all solutions of equation (5.3.1) is provided by the first-order linear equation

$$\frac{\mathrm{d}y}{\mathrm{d}t}-2ty=-t \tag{5.3.2}$$

The general solution of this equation is

$$y(t)=c\mathrm{e}^{t^2}+\frac{1}{2}$$

Now, observe that this solution is the sum of two terms; the first term, $c\mathrm{e}^{t^2}$, is the general solution of the homogeneous equation

$$\frac{\mathrm{d}y}{\mathrm{d}t}-2ty=0 \tag{5.3.3}$$

while the second term, $\frac{1}{2}$, is a solution of the non-homogeneous equation (5.3.2). In other words, every solution $y(t)$ of equation (5.3.2) is the sum of a particular solution, $\psi(t)=\frac{1}{2}$, with a solution ce^{t^2} of the homogeneous equation. A similar situation prevails in the case of the second-order equations, as we now show.

Theorem 5.3.1. Let $y_1(t)$ and $y_2(t)$ be two linearly independent solution of the homogeneous equation

$$L[y]=\frac{d^2y}{dt^2}+p(t)\frac{dy}{dt}+q(t)y=0 \tag{5.3.4}$$

and let $\psi(t)$ be any particular solution of the non-homogeneous equation (5.3.1).

Then, every solution $y(t)$ of equation (5.3.1) must be of the form

$$y(t)=c_1y_1(t)+c_2y_2(t)+\psi(t)$$

for some choice of constants c_1, c_2.

The proof of Theorem 5.3.1 relies heavily on the following lemma.

Lemma 5.3.1. The difference of any two solutions of the non-homogeneous equation (5.3.1) is a solution of the homogeneous equation (5.3.4).

Proof. Let $\psi_1(t)$ and $\psi_2(t)$ be two solutions of equation (5.3.1). By the linearity of L,

$$L[\psi_1-\psi_2](t)=L[\psi_1](t)-L[\psi_2](t)=g(t)-g(t)=0$$

Hence, $\psi_1(t)-\psi_2(t)$ is a solution of the homogeneous equation (5.3.4).

We can now give a very simple proof of Theorem 5.3.1.

Proof of Theorem 5.3.1. Let $y(t)$ be any solution of equation (5.3.1). By Lemma 5.3.1, the function $\phi(t)=y(t)-\psi(t)$ is a solution of the homogeneous equation (5.3.4). But every solution $\phi(t)$ of the homogeneous equation (5.3.4) is of the form $\phi(t)=c_1y_1(t)+c_2y_2(t)$, for some choice of constants c_1, c_2. Therefore,

$$y(t)=\phi(t)+\psi(t)=c_1y_1(t)+c_2y_2(t)+\psi(t)$$

Remark 5.3.1. Theorem 5.3.1 is an extremely useful theorem since it reduces the problem of finding all solutions of equation (5.3.4) to the much simple problem of finding just two solutions of the homogeneous equation (5.3.4), and one solution of the non-homogeneous equation (5.3.1).

【Example 5.3.1】 Find the general solution of the equation

$$\frac{d^2y}{dt^2}+y=t \tag{5.3.5}$$

Solution. The functions $y_1(t)=\cos t$ and $y_2(t)=\sin t$ are two linearly independent solutions of the homogeneous equation $y''+y=0$.

Moreover $\psi(t)=t$, is obviously a particular solution of equation (5.3.5). Therefore, by Theorem 5.3.1, every solution $y(t)$ of equation (5.3.5) must be of the form

$$y(t)=c_1\cos t+c_2\sin t+t$$

【Example 5.3.2】 Three solutions of a certain second-order non-homogeneous linear equation are $\psi_1(t)=t$, $\psi_2(t)=t+e^t$ and $\psi_3(t)=1+t+e^t$. Find the general solution of this equation.

Solution. By Lemma 5.3.1, the functions $\psi_2(t)-\psi_1(t)=e^t$ and $\psi_3(t)-\psi_2(t)=1$ are solutions of the corresponding homogeneous equation. Moreover, these functions are obviously linearly independent. Therefore, by Theorem 5.3.1, every solution $y(t)$ of this equation must be of the form

$$y(t)=c_1e^t+c_2+t$$

【Example 5.3.3】 Find a particular solution of $(D^2-D-2)y(x)=8e^{3x}$.

Solution. Since all derivatives of $8e^{3x}$ are of the form Ae^{3x}, we try $y_p=Ae^{3x}$. Substituting into the DE yields

$$9Ae^{3x}-3Ae^{3x}-2Ae^{3x}=8Ae^{3x}$$

Dividing by e^{3x}, we obtain $4A=8$, or $A=2$. This proves that if the DE has a particular solution of the form $y_p=Ae^{3x}$, then A must be 2. It is easy verified that $y_p=2e^{3x}$ does indeed define a particular solution of $y''-y'-2y=8e^{3x}$.

【Example 5.3.4】 Find a particular solution of

$$(D^2-7D+12)y(x)=12x^2+10x-11$$

Solution. Since differentiation reduces the degree of a polynomial, it is natural to assume that $y_p(x)$ is a polynomial of degree n whenever $\rho(x)$ is a polynomial of degree n.

Trying $y_p=Ax^2+Bx+C$ yields

$$(2A)-7(2Ax+B)+12(Ax^2+Bx+C)=12x^2+10x-11$$

Equating the coefficients of like powers of x, we obtain the equations

$$2A-7B+12C=-11$$
$$-14A+12B=10$$
$$12A=12$$

From these equations we find that $A=1$, $B=2$ and $C=1/12$. It is easily shown that $y_p=x^2+2x+1/12$ defines a particular solution.

【Example 5.3.5】 Find a particular solution of

$$(D^2+2D)y(x)=80\sin 4x$$

Solution. Since all derivatives of $80\sin 4x$ are of the form $A\sin 4x$ or $B\cos 4x$, we try

$$y_p=A\sin 4x+B\cos 4x$$

Substitution into the DE yields

$$(-16A\sin 4x-16B\cos 4x)+2(4A\cos 4x-4B\sin 4x)=80\sin 4x$$

Setting $4x=0$ and then $4x=\pi/2$, we obtain the equations

$$-16B+8A=0$$
$$-16A-8B=80$$

from which we obtain $A=-4$and $B=-2$.

It is easily shown that $y_p=-4\sin 4x-2\cos 4x$ define a particular solution of the given DE.

【Example 5. 3. 6】 Find a particular solution of

$$(D^2-3D)y(x)=2e^{3x}$$

Solution. Since $y_p=Ae^{3x}$ is part of the complete solution of $(D^2-3D)y(x)=0$ given by $y=c_1+c_2e^{3x}$, the given DE cannot have a particular solution of the form $y_p=Ae^{3x}$. Hence we try $y_p=Axe^{3x}$. Then

$$y'_p=3Axe^{3x}+Ae^{3x} \text{ and } y''_p=9Axe^{3x}+3Ae^{3x}+3Ae^{3x}$$

Substituting into $(D^2-3D)\ y\ (x)\ =2e^{3x}$, we obtain

$$(9Axe^{3x}+6Ae^{3x})-3(3Axe^{3x}+Ae^{3x})=2e^{3x}$$

And we find that $A=2/3$. We then verify that $y_p=\frac{2}{3}xe^{3x}$ defines a particular solution of $(D^2-3D)y(x)=2e^{3x}$.

More generally, if the characteristic equation of $Ly(x)=0$ has a root $r=\alpha+i\beta$ of multiplicity k, and $Y=\psi(x)$ defines the usual trial solution resulting from r, we try to find a particular solution of $Ly=\rho$ of the form $y_p=x^k\psi\ (x)$.

【Example 5. 3. 7】 Find a particular solution of

$$(D^2-3D+2)y(x)=12$$

Solution. Instead of trying $y_p=A$, we merely note by inspection that $y_p=6$ defines a particular solution.

【Example 5. 3. 8】 Find a complete solution of

$$(D^2-D-2)y(x)=8e^{3x}$$

Solution. From $r^2-r-2=0$, we obtain $r_1=2$ and $r_2=-1$. Hence, a complete solution of the reduces equation is given by $y_c=c_1e^{2x}+c_2e^{-x}$, often termed the complementary function. In Example 5. 3. 3, the given DE was found to possess the particular solution given by $y_p=2e^{3x}$. Hence, $y=c_1e^{2x}+c_2e^{-x}+2e^{3x}$ defines a complete solution. That is, every solution of $(D^2-D-2)y(x)=8e^{3x}$ can be expressed in the form

$$y=c_1e^{2x}+c_2e^{-x}+2e^{3x}$$

【Example 5. 3. 9】 Find a complete solution of $(D^2-3D)y(x)=2e^{3x}$ satisfying the conditions $y=1$ and $y'=3$ when $x=0$.

Solution. It follows from Example 5. 3. 6 that the given DE has a complete solution given by $y=c_1+c_2e^{3x}+\frac{2}{3}xe^{3x}$. Differentiating yields

$$y'=3c_2e^{2x}+2xe^{3x}+\frac{2}{3}e^{3x}$$

Using the given conditions, we obtain the equation $1=c_1+c_2$ and $3=3c_2+\frac{2}{3}$, and hence $c_2=\frac{7}{9}$ and $c_1=\frac{2}{9}$. Thus the required particular solution is given by

$$y=\frac{2}{9}+\frac{7}{9}e^{3x}+\frac{2}{3}xe^{3x}$$

【Example 5. 3. 10】 Find a complete solution of $(D^2-1)y(x)=5e^x\sin x$.

Solution. From $r^2-1=0$, we find $r=\pm 1$, and hence the complementary function is given by $y_c(x)=c_1e^x+c_2e^{-x}$.

The derivatives of $5e^x\sin x$ involve $e^x\sin x$ and $e^x\cos x$. Hence we try

$$y_p(x)=\psi(x)Ae^x\sin x+Be^x\cos x$$

Substituting

$$Dy_p(x)=Ae^x\cos x+Ae^x\sin x-Be^x\sin x+Be^x\cos x$$

and

$$D^2y_p(x)=-Ae^x\sin x+Ae^x\cos x+Ae^x\cos x+Ae^x\sin x$$
$$=-Be^x\cos x-Be^x\sin x+Be^x\cos x-Be^x\sin x$$

Into the given DE, we obtain

$$(2Ae^x\cos x-2Be^x\sin x)-(Ae^x\sin x+Be^x\cos x)=5e^x\sin x$$

Setting $x=0$ and $x=\pi/2$ yields the equations

$$-2A-B=0 \quad \text{and} \quad e^{\pi/2}(-2B-A)=5e^{\pi/2}$$

From which we find that $A=-1$ and $B=-2$. Thus

$$y_p(x)=-e^x\sin x-2e^x\cos x$$

It is a simple exercise to verify that

$$y(x)=y_c(x)+y_p(x)=c_1e^x+c_2e^{-x}-e^x\sin x-2e^x\cos x$$

defines a complete solution.

5.4 Higher order differential equations

In this section we briefly discuss higher-order linear differential equations.

Definition 5. 4. 1. The equation

$$L[y]=a_n(t)\frac{d^ny}{dt^n}+a_{n-1}(t)\frac{d^{n-1}y}{dt^{n-1}}+\cdots+a_0(t)y=0, a_n(t)\neq 0 \quad (5.4.1)$$

is called the general nth order homogeneous linear equation. The differential equation (5.4.1) together with the initial conditions

$$y(t_0)=y_0, y'(t_0)=y'_0, \cdots, y^{(n-1)}(t_0)=y_0^{(n-1)} \tag{5.4.2}$$

is called an initial-value problem. The theory for equation (5.4.1) is completely analogous to the theory for second-order linear homogeneous equation. Therefore, we will state the relevant theorems without proof.

Theorem 5.4.1. Let $y_1(t), \cdots, y_n(t)$ be n independent solutions of equation (5.4.1); that is, no solution $y_j(t)$ is a linear combination of the other solutions. Then, every solution $y(t)$ of equation (5.4.1) is of the form

$$y(t)=c_1 y_1(t)+c_2 y_2(t)+\cdots+c_n y_n(t) \tag{5.4.3}$$

for some choice of constants $c_1, c_2, \cdots, c_n$. For this reason, we say that equation (5.4.2) is the general solution of equation (5.4.1).

To find n independent solutions of equation (5.4.1) when the coefficients $a_0, a_1, \cdots, a_n$ do not depend on t, we compute

$$L[e^{rt}]=(a_n r^n+a_{n-1}r^{n-1}+\cdots+a_0)e^{rt} \tag{5.4.4}$$

This implies that e^{rt} is a solution of equation (5.4.1) if, and only if, r is a root of the characteristic equation

$$a_n r^n+a_{n-1}r^{n-1}+\cdots+a_0=0 \tag{5.4.5}$$

Thus, if equation (5.4.4) has n distinct roots $r_1, r_2, \cdots, r_n$, then the general solution of equation (5.4.1) is $y(t)=c_1 e^{r_1 t}+c_2 e^{r_2 t}+\cdots+c_n e^{r_n t}$ if $r_j=\alpha_j+i\beta_j$ is a complex root of equation (5.4.4), then

$$u(t)=\mathrm{Re}\{e^{\alpha_j t}\}=e^{\alpha_j t}\cos\beta_j t \tag{5.4.6}$$

and

$$v(t)=\mathrm{Im}\{e^{r_j t}\}=e^{\alpha_j t}\sin\beta_j t \tag{5.4.7}$$

are two real-valued solutions of equation (5.4.1). Finally, if r_1 is a root of multiplicity k; that is, if

$$a_n r^n+\cdots+a_0=(r-r_1)^k q(r)$$

where $q(r_1)=0$, then $e^{r_1 t}, te^{r_1 t}, \cdots, t^{k-1}e^{r_1 t}$ are k independent solutions of equation (5.4.1). We prove this last assertion in the following manner. Observe from equation (5.4.3) that

$$L[e^{rt}]=(r-r_1)^k q(r)e^{rt}$$

if r_1 is a root of multiplicity k. Therefore,

$$L[t^j e^{r_1 t}]=L\left[\frac{\partial^j}{\partial r^j}e^{rt}\right]\bigg|_{r=r_1}=\frac{\partial^j}{\partial r^j}L[e^{rt}]\bigg|_{r=r_1}$$

$$=\frac{\partial^j}{\partial r^j}(r-r_1)^k q(r)e^{rt}\bigg|_{r=r_1}=0, \text{for } 1\leqslant j\leqslant k$$

【Example 5.4.1】 Find the general solution of the equation

$$\frac{d^4 y}{dt^4}+y=0 \tag{5.4.8}$$

Solution. The characteristic equation of (5.4.8) is $r^4+1=0$. We find the roots of this equation by noting that $-1=e^{i\pi}=e^{3\pi i}=e^{5\pi i}=e^{7\pi i}$. Hence,

$$r_1=e^{i\pi/4}=\cos\frac{\pi}{4}+i\sin\frac{\pi}{4}=\frac{1}{\sqrt{2}}(1+i)$$

$$r_2=e^{3\pi i/4}=\cos\frac{3\pi}{4}+i\sin\frac{3\pi}{4}=-\frac{1}{\sqrt{2}}(1-i)$$

$$r_3=e^{5\pi i/4}=\cos\frac{5\pi}{4}+i\sin\frac{5\pi}{4}=-\frac{1}{\sqrt{2}}(1+i)$$

and

$r_4=e^{7\pi i/4}=\cos\frac{7\pi}{4}+i\sin\frac{7\pi}{4}=\frac{1}{\sqrt{2}}(1-i)$, are 4 roots of the equation $r^4+1=0$. The roots r_3 and r_4 are the complex conjugates of r_2 and r_1, respectively. Thus,

$$e^{r_1 t}=e^{t/\sqrt{2}}[\cos\frac{t}{\sqrt{2}}+i\sin\frac{t}{\sqrt{2}}]$$

and

$e^{r_2 t}=e^{-t/\sqrt{2}}[\cos\frac{t}{\sqrt{2}}+i\sin\frac{t}{\sqrt{2}}]$ are 2 complex-valued solutions of equation (5.4.8), and this implies that

$$y_1(t)=e^{t/\sqrt{2}}\cos\frac{t}{\sqrt{2}},\ y_2(t)=e^{t/\sqrt{2}}\sin\frac{t}{\sqrt{2}},\ y_3(t)=e^{-t/\sqrt{2}}\cos\frac{t}{\sqrt{2}},\ y_4(t)=e^{-t/\sqrt{2}}\sin\frac{t}{\sqrt{2}}$$

are 4 real-valued solutions of equation (5.4.8). These solutions are clearly independent.

Hence, the general solution of equation (5.4.8) is

$$y(t)=e^{t/\sqrt{2}}[a_1\cos\frac{t}{\sqrt{2}}+b_1\sin\frac{t}{\sqrt{2}}]+e^{-t/\sqrt{2}}[a_2\cos\frac{t}{\sqrt{2}}+b_2\sin\frac{t}{\sqrt{2}}].$$

【Example 5.4.2】 Find the general solution of the equation

$$\frac{d^4 y}{dt^4}-3\frac{d^3 y}{dt^3}+3\frac{d^2 y}{dt^2}-\frac{dy}{dt}=0 \tag{5.4.9}$$

Solution. The characteristic equation of (5.4.9) is

$$0=r^4-3r^3-r=r(r^3-3r^2+3r-1)=r(r-1)^3$$

Its roots are $r_1=0$ and $r_2=1$, with $r_2=1$ a root of multiplicity three. Hence, the general solution of equation (5.4.9) is

$$y(t)=c_1+(c_2+c_3 t+c_4 t^2)e^t$$

The theory for the non-homogeneous equation

$$L[y]=a_n(t)\frac{d^n y}{dt^n}+\cdots+a_0(t)y=f(t), a_n(t)\neq 0 \tag{5.4.10}$$

is also completely analogous to the theory for the second-order non-homogeneous equation.

Lemma 5.4.1. The difference of any two solutions of the non-homogeneous equation (5.4.10), and let $y_1(t), y_2(t), \cdots, y_n(t)$ be n independent solutions of the homogeneous equation (5.4.3) is of the form

$$y(t)=\psi(t)+c_1y_1(t)+\cdots+c_ny_n(t)$$

for some choice of constants $c_1, c_2, \cdots, c_n$.

The method of judicious guessing also applies to the nth-order equation

$$a_n\frac{d^n y}{dt^n}+a_{n-1}\frac{d^{n-1} y}{dt^n}+\cdots+a_0y=[b_0+b_1t+\cdots+b_kt^k]e^{at} \tag{5.4.11}$$

It is easily verified that equation (5.4.11) has a particular solution $\psi(t)$ of the form

$$\psi(t)=(c_0+c_1t+\cdots+c_kt^k)e^{at}$$

if e^{at} is not a solution of the homogeneous equation, and

$$\psi(t)=t^j(c_0+c_1(t)+\cdots+c_kt^k)e^{at}$$

if $t^{j-1}e^{at}$ is a solution of the homogeneous equation, but $t^j e^{at}$ is not.

【Example 5.4.3】 Find a particular solution $\psi(t)$ of the equation

$$L[y]=\frac{d^3 y}{dt^3+3\frac{d^2 y}{dt^2}+3\frac{dy}{dt}}+y=e^t \tag{5.4.12}$$

Solution. The characteristic equation

$$r^3+3r^2 3r+1=(r+1)^3$$

has $r=-1$ as a triple root. Hence, e^t is not a solution of the homogeneous equation, and equation (5.4.12) has a particular solution $\psi(t)$ of the form $\psi(t)=Ae^t$.

Computing $L[\psi](t)=8Ae^t$, we see that $A=1/8$.

Consequently,

$$\psi(t)=1/8e^t$$

is a particular solution of equation (5.4.12).

There is also a variation of parameters formula for the non-homogeneous equation (5.4.10). Let $v(t)$ be the solution of the homogeneous equation (5.4.4) which satisfies the initial conditions $v(t_0)=0, v'(t_0)=0, \cdots, v^{(n-2)}(t_0)=0, v^{(n-1)}(t_0)=1$ Then,

$$\psi(t)=\int_{t_0}^{t}\frac{v(t-s)}{a_n(s)}f(s)ds$$

is a particular solution of the non-homogeneous equation (5.4.10).

【Example 5.4.4】 It is easily verified that the function defined by $y=x^3-3$ is a solution on $(0,+\infty)$ of the initial-value problem

$$(D^3-x^{-1}D^2-3)y(x)=9;y(1)=-2,y'(1)=3,y''(1)=6$$

By Theorem 5.1.1, this solution is unique.

The principle of superposition applies almost immediately to the reduced equation $Ly(x)=0$ of DE (5.4.1) from the fact that L is a linear operator.

【Example 5.4.5】 The DE $(D^3-2D^2-D+2)y(x)=0$ has solution defined by $y_1=\mathrm{e}^x, y_2=\mathrm{e}^{-x}, y_3=\mathrm{e}^{2x}$. Hence $y_4=c_1\mathrm{e}^x+c_2\mathrm{e}^{-x}+c_3\mathrm{e}^{2x}$ also defines a solution.

Linear dependence was defined as follows:

Definition 5.4.2. A set of n function $y_1, y_2, \cdots, y_n$ that are not linearly independent on an interval I are said to be linearly dependent on I.

Lemma 5.4.1 generalizes to the following:

Theorem 5.4.2. There exist n linearly independent solutions of the reduced equation $Ly(x)=0$. These solutions are valid on any interval I on which $p_1, p_2, \cdots, p_n$ are continuous.

The Wronskian W of n functions $y_1, y_2, \cdots, y_n$ each of which can be differentiated $(n-1)$ times on an interval I, is defined on I by the determinant

$$W=W(y_1,y_2,\cdots,y_n)=\begin{vmatrix} y_1 & y_2 & y_3 & \cdots & y_n \\ y'_1 & y'_2 & y'_3 & \cdots & y'_n \\ y''_1 & y''_2 & y''_3 & \cdots & y''_n \\ \vdots & \vdots & \vdots & \vdots & \vdots \\ y_1^{(n-1)} & y_2^{(n-1)} & y_3^{(n-1)} & \cdots & y_n^{(n-1)} \end{vmatrix} \tag{5.4.13}$$

For n linearly independent solutions $y_1, y_2, \cdots, y_n$ of $Ly(x)=0$ on an interval I, Abel's identity generalizes to

$$W=W(x)=W(x_0)\exp-\left[\int_{x_0}^{x} p(t)\mathrm{d}t\right] \tag{5.4.14}$$

As in the case $n=2$, x_0 is an arbitrary point of I and W is everywhere positive, everywhere negative, or identically zero on I.

The Wronskian W can be obtained by employing any of the standard methods for evaluating a determinant. When $n=3$,

$$W=\begin{vmatrix} y_1 & y_2 & y_3 \\ y'_1 & y'_2 & y'_3 \\ y''_1 & y''_2 & y''_3 \end{vmatrix}=y_1\begin{vmatrix} y'_1 & y'_2 \\ y''_1 & y''_2 \end{vmatrix}-y'_1\begin{vmatrix} y_2 & y_3 \\ y''_2 & y''_3 \end{vmatrix}+y''_1\begin{vmatrix} y_2 & y_3 \\ y''_2 & y_3 \end{vmatrix}$$

$$=y_1(y'_2y''_3-y''_2y'_3)-y'_1(y_2y''_3-y''_2y_3)+y''_1(y_2y'_3-y'_2y_3)$$

$$=y_1y'_2y''_3+y_2y'_3y''_1+y''_3y'_1y''_2-y_3y'_2y''_1-y_2y'_1y''_3-y_1y'_3y''_2$$

【Example 5.4.6】 Find the Wronskian of the function defined by $y_1=1, y_2=x$,

$y_3=x^2$.

$$W(x)=\begin{vmatrix}1 & x & x^2\\0 & 1 & 2x\\0 & 0 & 2\end{vmatrix}=1\begin{vmatrix}1 & 2x\\0 & 2\end{vmatrix}-(0)\begin{vmatrix}x & x^2\\0 & 2\end{vmatrix}+(0)\begin{vmatrix}x & x^2\\1 & 2x\end{vmatrix}=2$$

Theorem 5.4.3. If each of $y_1, y_2, \cdots, y_n$ can be differentiated $(n-1)$ times on an interval I, and $W(y_1, y_2, \cdots, y_n)$ is not identically zero on I, then $y_1, y_2, \cdots, y_n$ vare linearly independent on I.

【Example 5.4.7】 The function of Example 5.4.6, given by $y_1=1, y_2=x, y_3=x^2$, are linearly-independent on an interval I since $W=2$ is not identically zero on I.

【Example 5.4.8】 The functions given by $y_1=\mathrm{e}^x, y_2=\sin x, y_3=\cos x$ are linearly independent on an arbitrary interval I since

$$\begin{aligned}W(x)&=\begin{vmatrix}\mathrm{e}^x & \sin x & \cos x\\ \mathrm{e}^x & \cos x & -\sin x\\ \mathrm{e}^x & -\sin x & -\cos x\end{vmatrix}\\&=\mathrm{e}^x(-\cos^2 x-\sin^2 x)-\mathrm{e}^x(-\sin x\cos x+\sin x\cos x)\\&\quad+\mathrm{e}^x(-\sin^2 x-\cos^2 x)=-2\mathrm{e}^x<0\end{aligned}$$

is not identically zero on I.

Theorem 5.4.4. If $y_1, y_2, \cdots, y_n$ are solutions of $Ly(x)=0$ on a given interval I, and if $W(y_1, y_2, \cdots, y_n)=0$ on I, then $y_1, y_2, \cdots, y_n$ are linearly dependent on I, that is, they are not linearly independent on I.

Theorem 5.4.5. If $y_1, y_2, \cdots, y_n$ are any n linearly independent solutions of $Ly(x)=0$ on a given u=interval I, then every solution on I of $Ly(x)=0$ can be expresses in the form $y=c_1y_1+c_2y_2+\cdots+c_ny_n$ where $c_1, c_2, \cdots, c_n$ are arbitrary constants.

Theorem 5.4.6. If $y=c_1y_1+c_2y_2+\cdots+c_ny_n$ defines a complete solution of $Ly(x)=0$ on a given interval I, and $y_{\mathrm{p}}=\psi(x)$ defines any particulars solution of $Ly=\rho$ on I then every solution of $Ly=\rho$ on I can be expressed in the form

$$y=c_1y_1+c_2y_2+\cdots+c_ny_n+y_{\mathrm{p}}$$

【Example 5.4.9】 It is seen by inspection that the functions defined by $y_1=1$, $y_2=x, y_3=x^2$ satisfy $D^3y(x)=0$ on $I=(-\infty, +\infty)$。In Example 5.4.6 we found that y_1, y_2, y_3 are linearly independent on I. By

Theorem 5.4.7. $y=c_1(1)+c_2x+c_3x^2$ defines a complete solution on $D^3y(x)=\mathrm{e}^x$ on I.

The method of solving second-order linear DE with constant coefficients extends readily to DE of the form $Ly=(D^n+a_1D^{n-1}+a_2D^{n-2}+\cdots+a_{n-1}D+a_n)y=\rho$ where $a_1,a_2,\cdots,a_n$ are real constants. We find that $y=\mathrm{e}^{rx}$ defines a solution of $Ly(x)=0$ if r is a root of the characteristic equation

$$g(r)=r^n+a_1r^{n-1}+\cdots+a_{n-1}r+a_n=0 \tag{5.4.15}$$

For $n=2$, the two roots (not necessarily distinct) are easily found from the quadratic formula. For a larger n, we need some facts concerning the exactly n roots $r_1,r_2,\cdots,r_n$. Some of these roots may be repeated and some may not be real numbers.

【Example 5.4.10】 The equation $r^3-r=r(r-1)(r+1)=0$ has roots $0,-1$, and 1.

【Example 5.4.11】 The equation $r^3-r^2-8r=(r-2)^2(r+3)=0$ has roots $2,2$, and -3.

【Example 5.4.12】 The equation $r^4-1=(r^2-1)(r^2+1)=0$ has roots $1,-1$, i and $-$i , where $\mathrm{i}=\sqrt{-1}$.

【Example 5.4.13】 Find a complete solution of $(D^3-7D+6)y(x)=0$.

Solution. From $r^3-7r+6=(r-1)(r-2)(r+3)=0$, we obtain $r_1=1,r_2=2,r_3=-3$. A complete solution of the DE is given by

$$y=c_1\mathrm{e}^x+c_2\mathrm{e}^{2x}+c_3\mathrm{e}^{-3x}$$

【Example 5.4.14】 Find a complete solution of $(D^3-3D-2)y(x)=0$.

Solution. From $r^3-3r-2=(r-1)^2(r-2)=0$, we obtain $r_1=r_2=-1,r_3=2$. A complete solution of the DE is given by

$$y=c_1\mathrm{e}^{-x}+c_2x\mathrm{e}^{-2x}+c_3\mathrm{e}^{2x}$$

【Example 5.4.15】 Find a complete solution of $(D^3+D)y(x)=0$.

Solution. From $r^3+r=r(r^2-1)=0$, we obtain $r_1=0,r_2=0+\mathrm{i},r_3=0-\mathrm{i}$. A complete solution of the DE is given by

$$y=c_1+c_2\sin x+c_3\cos x$$

【Example 5.4.16】 Find a complete solution of $(D^4+8D^2+16)y(x)=0$.

Solution. From $r^4+8r^2+16=(r^2+4)^2=0$, we obtain $r_1=r_2=2\mathrm{i},r_3=r_4=-2\mathrm{i}$. A complete solution of the DE is given by

$$y=c_1\sin 2x+c_2\cos 2x+c_3x\sin 2x+c_4x\cos 2x$$

【Example 5.4.17】 Find a complete solution of $[(D+3)(D^2+2D+5)^3]y(x)=0$.

Solution. From $(r+3)(r^2+2r+5)^3=0$, we obtain $r_1=-3,r_2=r_3=r_4=-1+2\mathrm{i}$, $r_5=r_6=r_7=-1-2\mathrm{i}$. A complete solution of the DE is given by

$$y=c_1\mathrm{e}^{-3x}+\mathrm{e}^{-x}(c_2\sin 2x+c_3\cos 2x)+x\mathrm{e}^{-x}(c_4\sin 2x+c_5\cos 2x)+$$
$$x^2\mathrm{e}^{-x}(c_6\sin 2x+c_7\cos 2x).$$

【Example 5.4.18】 Solve the initial-value problem:

$$Ly(x)=(D^3+D)y(x)=20e^{-2x}, y(0)=y'(0)=y''(0)=0$$

Solution. From $r(r^2+1)=0$, we obtain $r_1=0, r_2=i, r_3=-i$. A complete solution of $Ly(x)=0$ is given by $y=c_1+c_2\sin x+c_3\cos x$. Trying $y_p=Ae^{-2x}$ in $Ly(x)=20e^{-2x}$, we obtain $-8Ae^{-2x}-2Ae^{-2x}=20Ae^{-2x}$, and $A=-2$. A complete solution of $Ly(x)=20e^{-2x}$ is given by $y=c_1+c_2\sin x+c_3\cos x-2e^{-2x}$.

From

$$y'=c_2\cos x-c_3\sin x+4e^{-2x}$$

$$y''-c_2\sin x-c_3\cos x-8e^{-2x}$$

and the initial conditions, we obtain

$$y(0)=c_1+c_3-2=0, y'(0)=c_2+4, y''(0)=-c_3-8$$

These three equation yields $c_1=10, c_2=-4, c_3=-8$. The initial-value problem has the unique solution given by $y=10-4\sin x-8\cos x-2e^{-2x}$.

【Example 5.4.19】 Find a solution of $(D+1)^3y(x)=x^3e^{-x}$.

Solution.

$$(D+1)^2y(x)=\frac{1}{D+1}(x^3e^{-x})=e^{-x}\int e^x(x^3e^{-x})dx=\frac{x^4e^{-x}}{4}$$

$$(D+1)y(x)=\frac{1}{D+1}\left(\frac{x^4e^{-x}}{4}\right)=e^{-x}\int e^x\left(\frac{x^4e^{-x}}{4}\right)dx=\left(\frac{x^5e^{-x}}{20}\right)$$

$$y(x)=\frac{1}{D+1}\left(\frac{x^5e^{-x}}{20}\right)=e^{-x}\int e^x\left(\frac{x^5e^{-x}}{20}\right)dx=\frac{x^6e^{-x}}{120}$$

5.5 The Euler equation

An ordinary linear DE

$$Ly(x)=[a_0(x)D^n+a_1(x)D^{n-1}+\cdots+a_{n-1}(x)D+a_n(x)]y(x)=f(x)$$

is generally quite difficult to solve. Finding even one solution of $Ly(x)=0$ can present a formidable challenge.

It is sometimes possible to reduce a variable coefficient DE to a DE with constant coefficients by employing a suitable change to variables. An effective substitution with this property is often difficult to find. A simple illustration is afforded by the Euler DE

$$Ly(x)=(x^nD_x^n+a_1x^{n-1}D_x^{n-1}+\cdots+a_{n-1}xD_x+a_n)y(x)=\rho(x) \quad (5.5.1)$$

where x and D_x have the same exponent in each term. Restricting ourselves to an interval on which $x>0$, we eliminate the variable x by the substitution

$$x=e^t \quad \text{or} \quad t=\ln x$$

Denoting d/dt by D and noting that $dt/dx=1/x$, we obtain

$$\frac{dy}{dx}=\frac{dy}{dt}\frac{dt}{dx}=\frac{dy}{dt}\frac{1}{x}=\frac{1}{x}Dy$$

$$\frac{d^2y}{dx^2}=\frac{d}{dx}\left(\frac{dy}{dt}\frac{1}{x}\right)=\frac{dy}{dt}\left(-\frac{1}{x^2}\right)+\frac{1}{x}\frac{d^2y}{dt^2}\frac{dt}{dx}$$

$$=\frac{1}{x^2}\frac{d^2y}{dt^2}-\frac{1}{x^2}\frac{dy}{dt}=\frac{1}{x^2}D(D-1)y$$

$$\frac{d^3y}{dx^3}=\frac{d}{dx}\left[\frac{1}{x^2}\left(\frac{d^2y}{dt^2}-\frac{dy}{dt}\right)\right] \tag{5.5.2}$$

$$=\frac{1}{x^2}\left(\frac{d^3y}{dt^3}-\frac{d^2y}{dt^2}\right)\frac{dt}{dx}+\left(\frac{d^2y}{dt^2}-\frac{dy}{dt}\right)\left(\frac{-2}{x^3}\right)$$

$$=\frac{1}{x^3}\left(\frac{d^3y}{dt^3}-3\frac{d^2y}{dt^2}+2\frac{dy}{dt}\right)=\frac{1}{x^3}D(D-1)(D-2)y$$

$$\vdots$$

$$\frac{d^ny}{dx^n}=\frac{1}{x^n}D(D-1)(D-2)\cdots(D-n+1)y$$

【Example 5.5.1】 Find a complete solution of $x^2y''+xy'=0$ on $(0,+\infty)$.

Solution. Substituting for y' and y'', from equation (5.5.2), we obtain

$$x^2x^{-2}D(D-1)y+xx^{-1}Dy=0$$

From which

$$[D(D-1)+D]y=D^2y=0$$

From which $y=c_1+c_2t$.

A complete solution is given by $y=c_1+c_2\ln x$.

【Example 5.5.2】 Find a complete solution of

$$x^3y'''+5x^2y''+2xy'-2y=0 \text{ on } (0,+\infty)$$

Solution. Substituting for y', y'' and y''', from equation (5.5.2), we obtain

$$x^3x^{-3}D(D-1)(D-2)y+5x^2x^{-2}D(D-1)y+2xx^{-1}Dy-2y=0$$

Simplifying,

$$(D-1)[D(D-2)+5D+2]y=(D-1)(D+1)(D+2)y=0$$

From which

$$(r-1)(r+1)(r+2)=0; r=1,-1,-2$$

and $y=c_1e^t+c_2e^{-1}+c_3e^{-2t}$.

Replacement of t by $\ln x$ yields

$$y(x)=c_1\exp(\ln x)+c_2\exp(-\ln x)+c_3\exp(-2\ln x)$$

or

$$y(x)=c_1x+c_2x^{-1}+c_3x^{-2}$$

This defines a complete solution since $W(x,x^{-1},x^{-2})\neq 0$ on $(0,+\infty)$.

5.6 Exercises

1. Let $L[y](t)=y''(t)-3ty'(t)+3y(t)$

   Compute

   (1) $L[e^t]$, (2) $L[\cos\sqrt{3}t]$, (3) $L[2e^t+4\cos\sqrt{3}t]$, (4) $L[t^2]$, (5) $L[5t^2]$, (6) $L[t]$, (7) $L[t^2+3t]$,

2. Compute the Wronskian of the following pairs of functions.

   (1) $\sin at$, $\cos bt$; (2) $\sin^2 t$, $1-\cos 2t$; (3) t, $t\ln t$; (4) e^{at}, $te^{at}1$;

   (5) t, $t\ln t$ (6) $e^{at}\sin bt$, $e^{at}\cos bt$.

3. Find the general solution of each of the following equations.

   (1) $\dfrac{d^2y}{dx^2}-y=0$, (2) $\dfrac{d^2y}{dt^2}-3\dfrac{dy}{dt}+y=0$.

4. Solve each of the following initial-value problems.

   (1) $\dfrac{d^2y}{dt^2}-3\dfrac{dy}{dt}-4y=0, y(0)=1, y'(0)=0$,

   (2) $5\dfrac{d^2y}{dt^2}+5\dfrac{dy}{dx}-y=0, y(0)=0, y'(0)=0$.

5. Three solutions of a certain second-order non-homogeneous linear equations are

$$\psi_1(t)=t^2, \psi_2(t)=t^2+e^{2t} \text{ and } \psi_3(t)=1+t^2+2e^{2t}.$$

6. Find the general solution of this equation.

   Three solutions of a second-order linear equation $L[y]=g(t)$ are

$$\psi_1(t)=3e^t+e^{t^2}, \psi_2(t)=7e^t+e^{t^2} \text{ and } \psi_3(t)=5e^t+e^{-t^3}+e^t.$$

   Find the solution of the initial-value problem

$$L[y]=g, y(0)=1, y'(0)=2$$

7. Find the general solution of each of the following equations.

   (1) $y'''-2y''-y'+2y=0$;

   (2) $y^{\text{iv}}-5y'''+6y''+4y'-8y=0$.

8. Solve the following initial-value problems.

   $y^{\text{iv}}+4y'''+14y''-20y'+25y=0, y(0)=y'(0)=y''(0)=0, y'''(0)=0$.

9. Find a particular solution of each of the following equations.

   (1) $y'''+y'=\tan t$;

   (2) $y'''-4y'=t+\cos t+2e^{-2t}$.

Chapter 6 Systems of differential equations

In this chapter we can see that in a wide range of practical applications, more complex mathematical models will be more than one system of differential equations, and with some simplified assumptions and appropriate transformations, the equations can be transformed into first-order linear differential equations. This chapter studies the theory of linear differential equations, which is very worthy of attention. To study linear differential equations, we introduce the notation of vectors and matrices and make extensive use of the results of linear algebra (vector space and matrix algebra). Many theories of differential equations can only be properly and fully explained by means of knowledge of linear algebra.

6.1 Existence and uniqueness theorem

6.1.1 Marks and definitions

A first-order differential system with n unknown functions $x_1, x_2, \cdots, x_n$ is of the form

$$\begin{cases} x'_1 = f_1(t, x_1, x_2, \cdots, x_n) \\ x'_2 = f_2(t, x_1, x_2, \cdots, x_n) \\ \qquad \cdots \\ x'_n = f_n(t, x_1, x_2, \cdots, x_n) \end{cases} \tag{6.1.1}$$

If each $f_i(t, x_1, x_2, \cdots, x_n)$ is linear with respect to x_j for all $1 \leqslant j \leqslant n$, then system (6.1.1) is called a linear differential system[1]; otherwise, it is a nonlinear dif-

[1] linear differential system 线性微分方程组

ferential system[1].

When a set of differentiable functions $x_1(t), x_2(t), \cdots, x_n(t), t \in (a, b)$ satisfies identity

$$x'_i(t) = f_i(t, x_1, x_2, \cdots, x_n), i = 1, 2, \cdots, n \tag{6.1.2}$$

they are a group of solution of (6.1.1) for $t \in (a, b)$.

In addition, we call the solution including independent arbitrary constant c_1, $c_2, \cdots, c_n$

$$\begin{cases} x_1 = \varphi_1(t, c_1, c_2, \cdots, c_n) \\ x_2 = \varphi_2(t, c_1, c_2, \cdots, c_n) \\ \cdots \\ x_n = \varphi_n(t, c_1, c_2, \cdots, c_n) \end{cases} \tag{6.1.3}$$

a general solution of (6.1.1).

Next, we focus on the special case: first order linear differential system

$$\begin{cases} x'_1(t) = a_{11}(t)x_1(t) + a_{12}(t)x_2(t) + \cdots + a_{1n}(t)x_n(t) + f_1(t) \\ x'_2(t) = a_{21}(t)x_1(t) + a_{22}(t)x_2(t) + \cdots + a_{2n}(t)x_n(t) + f_2(t) \\ \cdots \\ x'_n(t) = a_{n1}(t)x_1(t) + a_{n2}(t)x_2(t) + \cdots + a_{nn}(t)x_n(t) + f_n(t) \end{cases} \tag{6.1.4}$$

where functions $a_{ij}(t)$ and $f_i(t)$, $(i, j = 1, 2, \cdots, n)$ are continuous.

(6.1.4) also can be written as

$$\boldsymbol{x}'(t) = \boldsymbol{A}(t)\boldsymbol{x}(t) + \boldsymbol{f}(t) \tag{6.1.5}$$

where $n \times n$ coefficient matrix $\boldsymbol{A}(t)$, n-dimensional column vectors $\boldsymbol{x}(t)$, $\boldsymbol{x}'(t)$, $\boldsymbol{f}(t)$ are as follows

$$\boldsymbol{A}(t) = \begin{bmatrix} a_{11}(t) & \cdots & a_{1n}(t) \\ \vdots & & \vdots \\ a_{n1}(t) & \cdots & a_{nn}(t) \end{bmatrix}, \boldsymbol{x}(t) = \begin{bmatrix} x_1(t) \\ \vdots \\ x_n(t) \end{bmatrix}, \boldsymbol{x}'(t) = \begin{bmatrix} x'_1(t) \\ \vdots \\ x'_n(t) \end{bmatrix}, \boldsymbol{f}(t) = \begin{bmatrix} f_1(t) \\ \vdots \\ f_n(t) \end{bmatrix} \tag{6.1.6}$$

Sometimes for simplicity, we also write the column vector as a transpose of the row one, such as $\boldsymbol{x} = (x_1, x_2, \cdots, x_n)^{\mathrm{T}}$.

The following concepts are needed for describing our theories.

A matrix or vector becomes continuous over an interval $[a, b]$ if each of its elements is a continuous function on $[a, b]$.

An $n \times n$ matrix $\boldsymbol{B}(t)$ or n-dimensional column vector $\boldsymbol{u}(t)$

---

[1] nonlinear differential system 非线性微分方程组

$$\boldsymbol{B}(t)=\begin{bmatrix} b_{11}(t) & b_{12}(t) & \cdots & b_{1n}(t) \\ b_{21}(t) & b_{22}(t) & \cdots & b_{2n}(t) \\ \vdots & \vdots & & \vdots \\ b_{n1}(t) & b_{n2}(t) & \cdots & b_{nn}(t) \end{bmatrix}, \quad \boldsymbol{u}(t)=\begin{bmatrix} u_1(t) \\ u_2(t) \\ \vdots \\ u_n(t) \end{bmatrix}$$

are differentiable about $t \in [a,b]$, if each of its elements is differentiable on $[a,b]$, whose derivative are, respectively,

$$\boldsymbol{B}'(t)=\begin{bmatrix} b'_{11}(t) & b'_{12}(t) & \cdots & b'_{1n}(t) \\ b'_{21}(t) & b'_{22}(t) & \cdots & b'_{2n}(t) \\ \vdots & \vdots & & \vdots \\ b'_{n1}(t) & b'_{n2}(t) & \cdots & b'_{nn}(t) \end{bmatrix}, \quad \boldsymbol{u}'(t)=\begin{bmatrix} u'_1(t) \\ u'_2(t) \\ \vdots \\ u'_n(t) \end{bmatrix}$$

It is easy to get that the following equations hold for $n \times n$ matrices $\boldsymbol{A}(t)$, $\boldsymbol{B}(t)$ and n-dimensional vectors $\boldsymbol{u}(t)$, $\boldsymbol{v}(t)$.

(1) $[\boldsymbol{A}(t)+\boldsymbol{B}(t)]'=\boldsymbol{A}'(t)+\boldsymbol{B}'(t)$,

$[\boldsymbol{u}(t)+\boldsymbol{v}(t)]'=\boldsymbol{u}'(t)+\boldsymbol{v}'(t)$;

(2) $[\boldsymbol{A}(t)\boldsymbol{B}(t)]'=\boldsymbol{A}'(t)\boldsymbol{B}(t)+\boldsymbol{A}(t)\boldsymbol{B}'(t)$;

(3) $[\boldsymbol{A}(t)\boldsymbol{u}(t)]'=\boldsymbol{A}'(t)\boldsymbol{u}(t)+\boldsymbol{A}(t)\boldsymbol{u}'(t)$.

Similarly, a matrix $\boldsymbol{B}(t)$ or vector $\boldsymbol{u}(t)$ is called integrable in the interval $[a,b]$ when each of its elements is integrable. Also, their integrals are demonstrated by

$$\int_a^b \boldsymbol{B}(t)\,\mathrm{d}t=\begin{bmatrix} \int_a^b b_{11}(t)\,\mathrm{d}t & \int_a^b b_{12}(t)\,\mathrm{d}t & \cdots & \int_a^b b_{1n}(t)\,\mathrm{d}t \\ \int_a^b b_{21}(t)\,\mathrm{d}t & \int_a^b b_{22}(t)\,\mathrm{d}t & \cdots & \int_a^b b_{2n}(t)\,\mathrm{d}t \\ \vdots & \vdots & & \vdots \\ \int_a^b b_{n1}(t)\,\mathrm{d}t & \int_a^b b_{n2}(t)\,\mathrm{d}t & \cdots & \int_a^b b_{nn}(t)\,\mathrm{d}t \end{bmatrix}$$

$$\int_a^b \boldsymbol{u}(t)\,\mathrm{d}t=\begin{bmatrix} \int_a^b u_1(t)\,\mathrm{d}t \\ \int_a^b u_2(t)\,\mathrm{d}t \\ \vdots \\ \int_a^b u_n(t)\,\mathrm{d}t \end{bmatrix}$$

Next we present the definition of solution to (6.1.5).

Definition 6.1.1. Let $\boldsymbol{A}(t)$ and $\boldsymbol{f}(t)$ ($t \in [a,b]$) are continuous $n \times n$ matrix and n-dimensional vector, respectively. Then the vector $\boldsymbol{u}(t)$ ($t \in [\alpha,\beta] \subset [a,b]$) is solution of system (6.1.5) if it satisfies (6.1.5) and $\boldsymbol{u}(t)$ is continuous on $[\alpha,\beta]$.

Adding initial conditions

$$x_1(t_0)=\eta_1,\ x_2(t_0)=\eta_2,\cdots,x_n(t_0)=\eta_n \tag{6.1.7}$$

or

$$\boldsymbol{x}(t_0)=\boldsymbol{\eta} \tag{6.1.8}$$

where $\boldsymbol{\eta}=(\eta_1,\eta_2,\cdots,\eta_n)$ together with system (6.1.5) produces an initial value problem.

Definition 6.1.2. The solution of the initial value problem[1]

$$\boldsymbol{x}'=\boldsymbol{A}(t)\boldsymbol{x}+\boldsymbol{f}(t),\ \boldsymbol{x}(t_0)=\boldsymbol{\eta} \tag{6.1.9}$$

is a solution $\boldsymbol{u}(t)$ $(t\in[\alpha,\beta]\subset[a,b])$ of (6.1.5) with $t_0\in[\alpha,\beta]$ satisfying $\boldsymbol{u}(t_0)=\boldsymbol{\eta}$.

【Example 6.1.1】 Try to list the differential equations of currents I_1 and I_2 through L_1 and L_2 in Fig. 6.1.1.

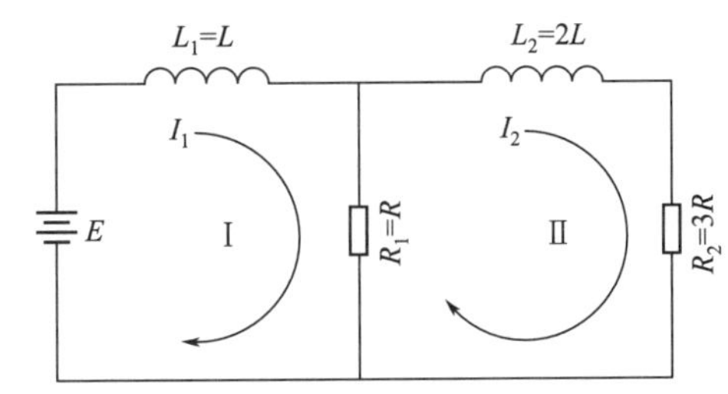

Figure 6.1.1.

Solution. Applying Kirchhoff's second law to loops Ⅰ and Ⅱ, the system of differential equations is obtained:

$$\begin{cases} L\dfrac{\mathrm{d}I_1}{\mathrm{d}t}+R(I_1-I_2)=E \\ 2L\dfrac{\mathrm{d}I_2}{\mathrm{d}t}+3RI_2+R(I_2-I_1)=0 \end{cases}$$

that is

$$\begin{cases} \dfrac{\mathrm{d}I_1}{\mathrm{d}t}=-\dfrac{R}{L}I_1+\dfrac{R}{L}I_2+\dfrac{E}{L} \\ \dfrac{\mathrm{d}I_2}{\mathrm{d}t}=\dfrac{R}{2L}I_1-\dfrac{2R}{L}I_2 \end{cases}$$

This is a first-order linear differential system about I_1 and I_2. Suppose

$$\boldsymbol{I}=\begin{bmatrix} I_1 \\ I_2 \end{bmatrix},\boldsymbol{A}=\begin{bmatrix} -\dfrac{R}{L} & \dfrac{R}{L} \\ \dfrac{R}{2L} & -\dfrac{2R}{L} \end{bmatrix},\ \boldsymbol{f}=\begin{bmatrix} \dfrac{E}{L} \\ 0 \end{bmatrix}$$

Then the system above can be written as $\boldsymbol{I}'=\boldsymbol{A}\boldsymbol{I}+\boldsymbol{f}$.

【Example 6.1.2】 Verify that the vector

$$\boldsymbol{u}(t)=\begin{bmatrix} \mathrm{e}^{-t} \\ -\mathrm{e}^{-t} \end{bmatrix}$$

[1] solution of the initial value problem 初值问题的解

is a solution with $t \in (-\infty, \infty)$ of the initial value problem

$$x' = \begin{bmatrix} 0 & 1 \\ 1 & 0 \end{bmatrix}, \quad x(0) = \begin{bmatrix} 1 \\ -1 \end{bmatrix}$$

Solution. Obviously,

$$u(0) = \begin{bmatrix} e^0 \\ -e^0 \end{bmatrix} = \begin{bmatrix} 1 \\ -1 \end{bmatrix}. e^{-t}$$

and $-e^{-t}$ have continuous derivatives everywhere, and

$$u'(t) = \begin{bmatrix} -e^{-t} \\ e^{-t} \end{bmatrix} = \begin{bmatrix} 0 & 1 \\ 1 & 0 \end{bmatrix} \begin{bmatrix} e^{-t} \\ -e^{-t} \end{bmatrix} = \begin{bmatrix} 0 & 1 \\ 1 & 0 \end{bmatrix} u(t)$$

so $u(t)$ is a solution of the initial value problem.

In addition, as seen in Chapter 4 & 5, to solve the initial value problem of nth order linear differential equation as $n \geqslant 2$ is difficult. It is further pointed out that the initial value problem of nth order linear differential equation can be solved by transferring into problem of linear differential system.

The initial value problem of nth order linear differential equation

$$\begin{cases} x^{(n)} + a_1(t)x^{(n-1)} + \cdots + a_{n-1}(t)x' + a_n(t)x = f(t) \\ x(t_0) = \eta_1, x'(t_0) = \eta_2, \cdots, x^{(n-1)}(t_0) = \eta_n \end{cases} \tag{6.1.10}$$

where $a_1(t), a_2(t), \cdots, a_n(t), f(t)$ are continuous functions over $[a, b]$, $t_0 \in [a, b]$, and $\eta_1, \eta_2, \cdots, \eta_n$ are known constants, could be converted into the following system of linear differential equations with initial value:

$$\begin{cases} x' = \begin{bmatrix} 0 & 1 & 0 & \cdots & 0 \\ 0 & 0 & 1 & \cdots & 0 \\ \vdots & \vdots & \vdots & & \vdots \\ 0 & 0 & 0 & \cdots & 1 \\ -a_n(t) & -a_{n-1}(t) & -a_{n-2}(t) & \cdots & -a_1(t) \end{bmatrix} x + \begin{bmatrix} 0 \\ 0 \\ \vdots \\ 0 \\ f(t) \end{bmatrix} \\ x(t_0) = \eta \end{cases} \tag{6.1.11}$$

where

$$x = \begin{bmatrix} x_1 \\ x_2 \\ \vdots \\ x_n \end{bmatrix}, x' = \begin{bmatrix} x'_1 \\ x'_2 \\ \vdots \\ x'_n \end{bmatrix}, \quad \eta = \begin{bmatrix} \eta_1 \\ \eta_2 \\ \vdots \\ \eta_n \end{bmatrix}$$

In fact, denote

$$x_1 = x, x_2 = x', x_3 = x'', \cdots, x_n = x^{(n-1)}$$

then

$$x'_1 = x' = x_2, x'_2 = x'' = x_3, \cdots, x'_{n-1} = x^{(n-1)} = x_n$$

$$x'_n = x^{(n)} = -a_n(t)x_1 - a_{n-1}(t)x_2 - \cdots - a_1(t)x_n + f(t)$$

with

$$x_1(t_0) = x(t_0) = \eta_1, x_2(t_0) = x'(t_0) = \eta_2, \cdots, x_n(t_0) = x^{(n-1)}(t_0) = \eta_n$$

Let $\psi(t)$ is any solution on $[a, b]$ containing t_0 of (6.1.10). Thus, $\psi(t)$, $\psi'(t)$, $\cdots$, $\psi^{(n)}(t)(t \in [a, b])$ are continuous and satisfy (6.1.10) with $\psi(t_0) = \eta_1$, $\psi'(t_0) = \eta_2$, $\cdots$, $\psi^{(n-1)}(t_0) = \eta_n$. Denote $\boldsymbol{\varphi}(t) = \begin{bmatrix} \varphi_1(t) \\ \varphi_2(t) \\ \vdots \\ \varphi_n(t) \end{bmatrix}$ where $\varphi_1(t) = \psi(t)$, $\varphi_2(t) = \psi'(t), \cdots, \varphi_n(t) = \psi^{(n-1)}(t)$. Certainly, $\boldsymbol{\varphi}(t_0) = \boldsymbol{\eta}$. Moreover,

$$\boldsymbol{\varphi}'(t) = \begin{bmatrix} \varphi'_1(t) \\ \varphi'_2(t) \\ \vdots \\ \varphi'_{n-1}(t) \\ \varphi'_n(t) \end{bmatrix} = \begin{bmatrix} \psi'(t) \\ \psi''(t) \\ \vdots \\ \psi^{(n-1)}(t) \\ \psi^{(n)}(t) \end{bmatrix}$$

$$= \begin{bmatrix} \varphi_2(t) \\ \varphi_3(t) \\ \vdots \\ \varphi_n(t) \\ -a_1(t)\psi^{(n-1)}(t) - \cdots - a_n(t)\psi(t) + f(t) \end{bmatrix}$$

$$= \begin{bmatrix} \varphi_2(t) \\ \varphi_3(t) \\ \vdots \\ \varphi_n(t) \\ -a_1(t)\varphi_n(t) - \cdots - a_n(t)\varphi_1(t) + f(t) \end{bmatrix}$$

$$= \begin{bmatrix} 0 & 1 & 0 & \cdots & 0 \\ 0 & 0 & 1 & \cdots & 0 \\ \vdots & \vdots & \vdots & & \vdots \\ 0 & 0 & 0 & \cdots & 1 \\ -a_n(t) & -a_{n-1}(t) & -a_{n-2}(t) & \cdots & -a_1(t) \end{bmatrix} \begin{bmatrix} \varphi_1(t) \\ \varphi_2(t) \\ \vdots \\ \varphi_{n-1}(t) \\ \varphi_n(t) \end{bmatrix} + \begin{bmatrix} 0 \\ 0 \\ \vdots \\ 0 \\ f(t) \end{bmatrix}$$

means $\boldsymbol{\varphi}(t)$ is a solution of (6.1.11). On the contrary, assume that the vector $\boldsymbol{u}(t) = (u_1(t), u_2(t), \cdots, u_n(t))$ is a solution of (6.1.11) with $t_0 \in [a, b]$ and define the function $w(t) = u_1(t)$, then from the ith equation of (6.1.11), $w^{(i)}(t) = u_i(t) = u_{i+1}(t)$, $i = 1, 2, \cdots, n-1$. Furthermore, the nth equation tells us

$$w^{(n)}(t)=u_n(t)$$
$$=-a_n(t)u_1(t)-a_{n-1}(t)u_2(t)-\cdots-a_2(t)u_{n-1}(t)-a_1(t)u_n(t)+f(t)$$
$$=-a_1(t)w^{(n-1)}(t)-a_2(t)w^{(n-2)}(t)-\cdots-a_n(t)w(t)+f(t)$$

that is

$$w^{(n)}(t)+a_1(t)w^{(n-1)}(t)+a_2(t)w^{(n-2)}(t)+\cdots+a_n(t)w(t)=f(t)$$

We can also get

$$w(t_0)=u_1(t_0)=\eta_1,\cdots,w^{(n-1)}(t_0)=u_n(t_0)=\eta_n$$

showing $w(t)$ is a solution of (6.1.10).

In summary, from the above discussion, we have proved that the initial value problems (6.1.10) and (6.1.11) are equivalent: given a solution of one initial value problems, we can construct a solution to another problem.

It is necessary to notice that each nth-order linear differential equation can be transformed into a system of n first-order linear differential equations, not vice versa. For example, the system of differential equations

$$\boldsymbol{x}'=\begin{bmatrix}1 & 0\\ 0 & 1\end{bmatrix}\boldsymbol{x},\boldsymbol{x}=\begin{bmatrix}x_1\\ x_2\end{bmatrix}$$

cannot be turned into a second-order differential equation.

6.1.2 Existence and uniqueness of solutions

This part is dedicated to studying existence and uniqueness theorem of solutions to the initial value problem (6.1.9).

We use the stepwise approximation method to prove the theorem through 5 small propositions. Because the systems are now discussed (written in the form of vectors), there is great complexity, in addition, norm of vectors and matrices, and convergence of vector function sequences are introduced; however, since the systems are linear, the situation is simpler and the conclusions are strengthened.

For $n\times n$ matrix $\boldsymbol{A}=[a_{ij}]_{n\times n}$ and n-dimensional vector $\boldsymbol{x}=(x_1,x_2,\cdots,x_n)^{\mathrm{T}}$, we give the definition of norm[1]

$$\|\boldsymbol{A}\|=\sum_{i,j=1}^{n}|a_{ij}|,\ \|\boldsymbol{x}\|=\sum_{i=1}^{n}|x_i|$$

Let $\boldsymbol{A}$, $\boldsymbol{B}$ are $n\times n$ matrices, and $\boldsymbol{x}$, $\boldsymbol{y}$ are n-dimensional vectors, then it is easy to check that

(1) $\|\boldsymbol{AB}\|\leqslant\|\boldsymbol{A}\|\cdot\|\boldsymbol{B}\|$,

$\|\boldsymbol{Ax}\|\leqslant\|\boldsymbol{A}\|\cdot\|\boldsymbol{x}\|$;

---

[1] norm 范数

(2) $\|\boldsymbol{A}+\boldsymbol{B}\| \leqslant \|\boldsymbol{A}\|+\|\boldsymbol{B}\|$,

$\|\boldsymbol{x}+\boldsymbol{y}\| \leqslant \|\boldsymbol{x}\|+\|\boldsymbol{y}\|$.

The vector sequence $\{\boldsymbol{x}_k\}$, $\boldsymbol{x}_k=(x_{1k},x_{2k},\cdots,x_{nk})^{\mathrm{T}}$ is convergent if the sequence $\{x_{ik}\}$ is convergent for each $i\,(i=1,2,\cdots,n)$.

The vector function sequence $\{\boldsymbol{x}_k(t)\}$, $\boldsymbol{x}_k(t)=(x_{1k}(t),x_{2k}(t),\cdots,x_{nk}(t))^{\mathrm{T}}$ is called convergent (uniformly convergent)[1] over $[a,b]$ if the function sequence $\{x_{ik}(t)\}$ is convergent (uniformly convergent) on $[a,b]$ for any i $(i=1,2,\cdots,n)$. It is easy to know that the uniformly convergent limit vector function of the sequence $\{\boldsymbol{x}_k(t)\}$ of continuous vector functions is still continuous.

Vector function series $\sum_{k=1}^{\infty}\boldsymbol{x}_k(t)$ is convergent (uniformly convergent) on $[a,b]$ if its sequence of partial sum is convergent (uniformly convergent). Also, the method of discriminating the uniform convergence of the usual function series is also true for the vector function series, that is, if

$$\|\boldsymbol{x}_k(t)\| \leqslant M_k,\ a\leqslant t\leqslant b,$$

and $\sum_{k=1}^{\infty}M_k$ is convergent, then $\sum_{k=1}^{\infty}\boldsymbol{x}_k(t)$ is uniformly convergent over $[a,b]$. Besides, provided that the sequence of vector functions $\{\boldsymbol{x}_k(t)\}$ is uniformly convergent $\lim\limits_{k\to\infty}\int_a^b \boldsymbol{x}_k(t)\,\mathrm{d}t=\int_a^b \lim\limits_{k\to\infty}\boldsymbol{x}_k(t)\,\mathrm{d}t$. For the general matrix sequence, similar definitions and results can be obtained like those above.

For instance, $n\times n$ matrix sequence $\{\boldsymbol{A}_k\}$ where $\boldsymbol{A}_k=[a_{ij}^{(k)}]_{n\times n}$ is convergent provided that sequence $\{a_{ij}\}$ is convergent for any $i,j=1,2,\cdots,n$.

Infinite matrix series

$$\sum_{k=1}^{\infty}\boldsymbol{A}_k=\boldsymbol{A}_1+\boldsymbol{A}_2+\cdots+\boldsymbol{A}_k+\cdots$$

is named convergent when its sequence of partial sums is convergent.

Similarly, for each integer k,

$$\|\boldsymbol{A}_k\| \leqslant M_k$$

and convergent $\sum_{k=1}^{\infty}M_k$ makes $\sum_{k=1}^{\infty}\boldsymbol{A}_k$ be convergent.

Also, the definition of the uniform convergence of the series of infinite matrix functions and the related results can be given.

Next, we have the main result about existence and uniqueness of solutions to problem (6.1.9).

---

[1] convergent (uniformly convergent) 收敛（一致收敛）

Theorem 6.1.1. (Existence and Uniqueness Theorem). If $n \times n$ matrix $\mathbf{A}(t)$ and n-dimensional column vector $\boldsymbol{f}(t)$ are continuous on a closed interval $[a,b]$, then there exists only one solution $\boldsymbol{\varphi}(t)$, $t \in [a,b]$ of problem (6.1.9) with $t_0 \in [a,b]$.

Theorem 6.1.1 can be proved through the following 5 propositions.

Proposition 6.1.1. If $\boldsymbol{\varphi}(t)$ ($t \in [a,b]$) is a solution of (6.1.9), then $\boldsymbol{\varphi}(t)$ is a continuous solution to the integral equation

$$\boldsymbol{x}(t)=\boldsymbol{\eta}+\int_{t_0}^{t} [\mathbf{A}(s)\boldsymbol{x}(s)+\boldsymbol{f}(s)]\mathrm{d}s, \ a \leqslant t \leqslant b \tag{6.1.12}$$

and vice verse.

Its proof is completely similar to the first-order differential equation, and we will not repeat it.

Let $\boldsymbol{\varphi}_0(t)=\boldsymbol{\eta}$, and construct the sequence of vector functions

$$\begin{cases} \boldsymbol{\varphi}_0(t)=\boldsymbol{\eta}, & a \leqslant t \leqslant b \\ \boldsymbol{\varphi}_k(t)=\boldsymbol{\eta}+\int_{t_0}^{t} [\mathbf{A}(s)\boldsymbol{\varphi}_{k-1}(s)+\boldsymbol{f}(s)]\mathrm{d}s, & k=1,2,\cdots \end{cases} \tag{6.1.13}$$

The vector function $\boldsymbol{\varphi}_k(t)$ is the kth approximate solution[1] to (6.1.9). Mathematical induction immediately pushes Proposition 6.1.2.

Proposition 6.1.2. For each positive integer k, the vector function $\boldsymbol{\varphi}_k(t)$ is defined and continuous over the interval $[a,b]$.

Proposition 6.1.3. The sequence of vector functions $\{\boldsymbol{\varphi}_k(t)\}$ is uniformly convergent on $[a,b]$.

Proof. With regard to the vector function series

$$\boldsymbol{\varphi}_0(t)+\sum_{j=1}^{\infty} [\boldsymbol{\varphi}_j(t)-\boldsymbol{\varphi}_{j-1}(t)], \ a \leqslant t \leqslant b \tag{6.1.14}$$

The partial sum of series (6.1.14) is

$$\boldsymbol{\varphi}_0(t)+\sum_{j=1}^{k} [\boldsymbol{\varphi}_j(t)-\boldsymbol{\varphi}_{j-1}(t)]=\boldsymbol{\varphi}_k(t)$$

so we need to prove that series (6.1.14) is uniformly convergent in order to get uniform convergence of sequence $\{\boldsymbol{\varphi}_k(t)\}$. Since both $\mathbf{A}(t)$ and $\boldsymbol{f}(t)$ are continuous over $[a,b]$, $\mathbf{A}(t)$ and $\boldsymbol{f}(t)$ are bounded. Suppose L and K are positive constants such that

$$\|\mathbf{A}(t)\| \leqslant L, \|\boldsymbol{f}(t)\| \leqslant K, \ a \leqslant t \leqslant b$$

and let $M=L\|\boldsymbol{\eta}\|+K$. We only prove that the sequence $\{\boldsymbol{\varphi}(t)\}$ converges uniformly over the interval $[t_0,b]$, and the uniform convergence on $[a,t_0]$ can be

---

[1] the kth approximate solution 第 k 次近似解

similarly proved. Based on (6.1.13), we give the following estimates

$$\begin{aligned}\|\boldsymbol{\varphi}_1(t)-\boldsymbol{\varphi}_0(t)\| &\leqslant \int_{t_0}^{t} \|\boldsymbol{A}(s)\boldsymbol{\varphi}_0(s)+\boldsymbol{f}(s)\|\,\mathrm{d}s \\ &\leqslant \int_{t_0}^{t} \|\boldsymbol{A}(s)\boldsymbol{\varphi}_0(s)\|+\|\boldsymbol{f}(s)\|\,\mathrm{d}s \\ &\leqslant \int_{t_0}^{t} [L\|\boldsymbol{\eta}\|+K]\,\mathrm{d}s = M(t-t_0)\end{aligned} \tag{6.1.15}$$

And

$$\|\boldsymbol{\varphi}_2(t)-\boldsymbol{\varphi}_1(t)\| \leqslant \int_{t_0}^{t} \|\boldsymbol{A}(s)[\boldsymbol{\varphi}_1(s)-\boldsymbol{\varphi}_0(s)]\|\,\mathrm{d}s$$

furthermore, it is obtained that from (6.1.15)

$$\|\boldsymbol{\varphi}_2(t)-\boldsymbol{\varphi}_1(t)\| \leqslant L\int_{t_0}^{t} M(s-t_0)\,\mathrm{d}s = \frac{ML}{2!}(t-t_0)^2$$

If

$$\|\boldsymbol{\varphi}_j(t)-\boldsymbol{\varphi}_{j-1}(t)\| \leqslant \frac{ML^{j-1}}{j!}(t-t_0)^j$$

is true, then (6.1.13) tells us that

$$\begin{aligned}\|\boldsymbol{\varphi}_{j+1}(t)-\boldsymbol{\varphi}_j(t)\| &\leqslant \int_{t_0}^{t} \|\boldsymbol{A}(s)[\boldsymbol{\varphi}_j(s)-\boldsymbol{\varphi}_{j-1}(s)]\|\,\mathrm{d}s \\ &\leqslant L\int_{t_0}^{t} \frac{ML^{j-1}}{j!}(s-t_0)^j\,\mathrm{d}s \leqslant \frac{ML^j}{(j+1)!}(t-t_0)^{j+1}\end{aligned}$$

Then we have the estimates for any positive integer k by mathematical induction

$$\|\boldsymbol{\varphi}_k(t)-\boldsymbol{\varphi}_{k-1}(t)\| \leqslant \frac{ML^{k-1}}{k!}(t-t_0)^k,\quad t_0\leqslant t\leqslant b \tag{6.1.16}$$

thus, as $t\in[t_0,b]$

$$\|\boldsymbol{\varphi}_k(t)-\boldsymbol{\varphi}_{k-1}(t)\| \leqslant \frac{ML^{k-1}}{k!}(b-t_0)^k \tag{6.1.17}$$

The right hand of (6.1.17) is the general term of positive term series

$$\frac{M}{L}\sum_{k=1}^{\infty}\frac{L^k(b-t_0)^k}{k!}$$

so from discriminant method for uniform convergence of vector function series, the series (6.1.14) converges uniformly over $[t_0,b]$. Finally, the sequence of vector functions $\{\boldsymbol{\varphi}_k(t)\}$ is also uniformly convergent on $[t_0,b]$.

Denoting

$$\lim_{k\to\infty}\boldsymbol{\varphi}_k(t)=\boldsymbol{\varphi}(t)$$

because $\boldsymbol{\varphi}(t)$ is the uniformly convergent limit function of $\boldsymbol{\varphi}_k(t)$, $\boldsymbol{\varphi}(t)$ is continuous on $[a,b]$.

Proposition 6. 1. 4. $\boldsymbol{\varphi}(t)$ is a continuous solution to integral equation (6. 1. 12) defined on $[a,b]$.

Proof. $\{\boldsymbol{\varphi}_k(t)\}$ uniformly converges to $\boldsymbol{\varphi}(t)$ over $[a,b]$ and $\mathbf{A}(t)$ is continuous, therefore the sequence $\{\mathbf{A}(s)\boldsymbol{\varphi}_k(t)\}$ uniformly converges to $\mathbf{A}(s)\boldsymbol{\varphi}(s)$ on $[a,b]$. Taking the limit on both sides of (6. 1. 13),

$$\begin{aligned}\lim_{k\to\infty}\boldsymbol{\varphi}_k(t)&=\boldsymbol{\eta}+\lim_{k\to\infty}\int_{t_0}^{t}[\mathbf{A}(s)\boldsymbol{\varphi}_{k-1}(s)+\boldsymbol{f}(s)]\,\mathrm{d}s\\&=\boldsymbol{\eta}+\int_{t_0}^{t}[\lim_{k\to\infty}\mathbf{A}(s)\boldsymbol{\varphi}_{k-1}(s)+\boldsymbol{f}(s)]\,\mathrm{d}s\end{aligned}$$

that is,

$$\boldsymbol{\varphi}(t)=\boldsymbol{\eta}+\int_{t_0}^{t}[\mathbf{A}(s)\boldsymbol{\varphi}(s)+\boldsymbol{f}(s)]\,\mathrm{d}s$$

Hence, $\boldsymbol{\varphi}(t)$ is a continuous solution over $[a,b]$ of integral equation (6. 1. 12).

Proposition 6. 1. 5. If $\boldsymbol{\psi}(t)$ is another continuous solution to integral equation (6. 1. 12) defined on $[a,b]$, then $\boldsymbol{\varphi}(t)=\boldsymbol{\psi}(t)$ $(t\in[a,b])$.

Proof. We firstly prove that $\boldsymbol{\psi}(t)$ is also the uniformly convergent limit function of the sequence $\{\boldsymbol{\varphi}_k(t)\}$ over $[a,b]$. According to (6. 1. 13) and

$$\boldsymbol{\psi}(t)=\boldsymbol{\eta}+\int_{t_0}^{t}[\mathbf{A}(s)\boldsymbol{\psi}(s)+\boldsymbol{f}(s)]\,\mathrm{d}s$$

the estimates can be given like Proposition 6. 1. 3

$$\|\boldsymbol{\varphi}_k(t)-\boldsymbol{\psi}(t)\|\leqslant\frac{\widetilde{M}L^k}{(k+1)!}(t-t_0)^{k+1},t_0\leqslant t\leqslant b$$

and then as $t\in[t_0,b]$, we have

$$\|\boldsymbol{\varphi}_k(t)-\boldsymbol{\psi}(t)\|\leqslant\frac{\widetilde{M}L^k}{(k+1)!}(b-t_0)^{k+1},t_0\leqslant t\leqslant b$$

As we know, the series with the general term $\dfrac{\widetilde{M}L^k}{(k+1)!}(b-t_0)^{k+1}$ is convergent, as a result, $\dfrac{\widetilde{M}L^k}{(k+1)!}(b-t_0)^{k+1}\to 0$ $(k\to\infty)$, that is, $\boldsymbol{\varphi}_k(t)$ uniformly converges $\boldsymbol{\psi}(t)$ on $[t_0,b]$. In the light of uniqueness,

$$\boldsymbol{\varphi}(t)-\boldsymbol{\psi}(t),\ t_0\leqslant t\leqslant b$$

For $t\in[a,t_0]$, it can be proved similarly. The proof is completed.

Integrate Propositions 6. 1. 1~6. 1. 5, the proof about existence and uniqueness theorem is shown.

Because of the equivalence between the initial value problem (6. 1. 10) and (6. 1. 11) in 6. 1. 1, immediately, the existence and uniqueness theorem of n-order linear differential equations can be derived from Theorem 6. 1. 1.

Corollary 6.1.1. Assume that $a_1(t), a_2(t), \cdots, a_n(t), f(t)$ are continuous functions over $[a, b]$, then for each $t \in [a, b]$ and any $\eta_1(t), \eta_2(t), \cdots, \eta_n(t)$, the equation

$$x^{(n)} + a_1(t)x^{(n-1)} + \cdots + a_{n-1}(t)x' + a_n(t)x = f(t)$$

has a unique solution $w(t)$ $(t \in [a, b])$ satisfying the initial value condition

$$w(t_0) = \eta_1,\ w'(t_0) = \eta_2,\ \cdots,\ w^{(n-1)}(t_0) = \eta_n$$

6.2 General theory of linear differential systems

This section is devoted to the theory of linear differential system

$$\boldsymbol{x}' = \boldsymbol{A}(t)\boldsymbol{x} + \boldsymbol{f}(t) \tag{6.2.1}$$

mainly about the structure of solutions.

Indeed, (6.2.1) is called inhomogeneous if $\boldsymbol{f}(t) \neq \boldsymbol{0}$, where $\boldsymbol{0}$ is n-dimensional zero vector. If $\boldsymbol{f}(t) = \boldsymbol{0}$, then the equation

$$\boldsymbol{x}' = \boldsymbol{A}(t)\boldsymbol{x} \tag{6.2.2}$$

is homogeneous[1], and is called a homogeneous linear differential equation system[2] corresponding to (6.2.1).

6.2.1 Linear homogeneous systems

This section mainly studies the algebraic structure of solutions to homogeneous linear differential equations (6.2.2). We assume that the matrix $\boldsymbol{A}(t)$ is continuous on $[a, b]$.

Let $\boldsymbol{u}(t)$ and $\boldsymbol{v}(t)$ be arbitrary two solutions of (6.2.2), then according to the differential law of vector functions, $\alpha\boldsymbol{u}(t) + \beta\boldsymbol{v}(t)$ is also a solution to (6.2.2), where α and β are any constants. So we get superposition principle to solutions in linear homogeneous system.

Theorem 6.2.1. (Superposition Principle) Suppose that $\boldsymbol{u}(t)$ and $\boldsymbol{v}(t)$ are solutions to (6.2.2), then the linear combination $\alpha\boldsymbol{u}(t) + \beta\boldsymbol{v}(t)$ is a solution, too, where α and β are any constants.

Theorem 6.2.1 states that the set of all solutions of (6.2.2) constitutes a linear space. What is its dimension? To make sure that, we need the notion about linearly dependent or independent function vectors.

Definition 6.2.1. A group of n vector valued functions $\boldsymbol{x}_1(t), \boldsymbol{x}_2(t), \cdots, \boldsymbol{x}_n(t)$ is linearly dependent[3] on interval $[a, b]$ if there exist not all zero constants c_1, c_2,

---

[1] homogeneous 齐次

[2] homogeneous linear differential equation system 齐次线性微分方程组

[3] linearly dependent 线性相关

$\cdots, c_n$ such that

$$c_1 \boldsymbol{x}_1(t)+c_2 \boldsymbol{x}_2(t)+\cdots+c_n \boldsymbol{x}_n(t)=\mathbf{0}$$

Otherwise, the set of vectors is called linearly independent[1] on $[a, b]$.

【Example 6. 2. 1】 Show that vector valued functions

$$\boldsymbol{x}_1(t)=\begin{bmatrix}\cos^2 t\\ 1\\ t\end{bmatrix},\ \boldsymbol{x}_2(t)=\begin{bmatrix}-\sin^2 t+1\\ 1\\ t\end{bmatrix}$$

are linearly dependent on any interval I.

Solution. Set $c_1=1$, $c_2=-1$, and we obtain that

$$c_1\begin{bmatrix}\cos^2 t\\ 1\\ t\end{bmatrix}+c_2\begin{bmatrix}-\sin^2 t+1\\ 1\\ t\end{bmatrix}=\begin{bmatrix}0\\ 0\\ 0\end{bmatrix}, t\in I$$

therefore, $\boldsymbol{x}_1(t)$, $\boldsymbol{x}_2(t)$ are linearly dependent on I.

【Example 6. 2. 2】 Show that vector valued functions

$$\boldsymbol{x}_1(t)=\begin{bmatrix}\mathrm{e}^{t}\\ 0\\ \mathrm{e}^{-t}\end{bmatrix}, \boldsymbol{x}_2(t)=\begin{bmatrix}0\\ \mathrm{e}^{3t}\\ 1\end{bmatrix}, \boldsymbol{x}_3(t)=\begin{bmatrix}\mathrm{e}^{2t}\\ \mathrm{e}^{3t}\\ 0\end{bmatrix}$$

are linearly independent on $(-\infty, +\infty)$.

Solution. In order to assure that

$$c_1 \boldsymbol{x}_1(t)+c_2 \boldsymbol{x}_2(t)+c_3 \boldsymbol{x}_3(t)=c_1\begin{bmatrix}\mathrm{e}^{t}\\ 0\\ \mathrm{e}^{-t}\end{bmatrix}+c_2\begin{bmatrix}0\\ e^{3t}\\ 1\end{bmatrix}+c_3\begin{bmatrix}\mathrm{e}^{2t}\\ \mathrm{e}^{3t}\\ 0\end{bmatrix}=\mathbf{0},\ t\in \boldsymbol{R}$$

that is,

$$\begin{bmatrix}\mathrm{e}^{t} & 0 & \mathrm{e}^{2t}\\ 0 & \mathrm{e}^{3t} & \mathrm{e}^{3t}\\ \mathrm{e}^{-t} & 1 & 0\end{bmatrix}\begin{bmatrix}c_1\\ c_2\\ c_3\end{bmatrix}=\begin{bmatrix}0\\ 0\\ 0\end{bmatrix},\ -\infty<t<+\infty$$

Simple calculation demonstrates that

$$\begin{vmatrix}\mathrm{e}^{t} & 0 & \mathrm{e}^{2t}\\ 0 & \mathrm{e}^{3t} & \mathrm{e}^{3t}\\ \mathrm{e}^{-t} & 1 & 0\end{vmatrix}=-2\mathrm{e}^{4t}<0$$

This gives us that $c_1=c_2=c_3=0$, that is to say, $\boldsymbol{x}_1(t)$, $\boldsymbol{x}_2(t)$, $\boldsymbol{x}_3(t)$ are linearly independent.

Example 6. 2. 2 shows a method to decide whether vector valued functions are linearly dependent or not. Define n-order determinant

---

[1] linearly independent 线性无关

$$W(t)=\begin{vmatrix} x_{11}(t) & x_{12}(t) & \cdots & x_{1n}(t) \\ x_{21}(t) & x_{22}(t) & \cdots & x_{2n}(t) \\ \vdots & \vdots & & \vdots \\ x_{n1}(t) & x_{n1}(t) & \cdots & x_{nn}(t) \end{vmatrix}$$

consisting of n-vector valued functions on $[a,b]$

$$\boldsymbol{x}_1(t)=\begin{bmatrix} x_{11}(t) \\ x_{21}(t) \\ \vdots \\ x_{n1}(t) \end{bmatrix}, \boldsymbol{x}_2(t)=\begin{bmatrix} x_{12}(t) \\ x_{22}(t) \\ \vdots \\ x_{n2}(t) \end{bmatrix}, \cdots, \boldsymbol{x}_n(t)=\begin{bmatrix} x_{1n}(t) \\ x_{2n}(t) \\ \vdots \\ x_{nn}(t) \end{bmatrix}$$

to be their Wronskian determinant.

Theorem 6.2.2. Suppose that vector functions $\boldsymbol{x}_1(t), \boldsymbol{x}_2(t), \cdots, \boldsymbol{x}_n(t)$ $(t \in [a,b])$ are linearly dependent, then their Wronskian determinant $W(t)=0$ $(t \in [a,b])$.

Proof. If $\boldsymbol{x}_1(t), \boldsymbol{x}_2(t), \cdots, \boldsymbol{x}_n(t)$ $(t \in [a,b])$ are linearly dependent, then there exist not all zero constants $c_1, c_2, \cdots, c_n$ such that

$$c_1\boldsymbol{x}_1(t)+c_2\boldsymbol{x}_2(t)+\cdots+c_n\boldsymbol{x}_n(t)=\boldsymbol{0},\ a \leqslant t \leqslant b \tag{6.2.3}$$

(6.2.3) is a linear homogeneous algebraic equations with unknown $c_1, c_2, \cdots, c_n$ whose coefficient determinant is Wronskian determinant $W(t)$ composed of $\boldsymbol{x}_1(t)$, $\boldsymbol{x}_2(t), \cdots, \boldsymbol{x}_n(t)$. That is, the system (6.2.3) has non-zero solution, then its coefficient determinant $W(t)=0$, $a \leqslant t \leqslant b$.

Theorem 6.2.3. If solutions $\boldsymbol{x}_1(t), \boldsymbol{x}_2(t), \cdots, \boldsymbol{x}_n(t)$ of (6.2.2) are linearly independent, then their Wronskian determinant $W(t) \neq 0$, $a \leqslant t \leqslant b$.

Proof. Applying proof by contradiction, assume that there is a t_0 $(a \leqslant t_0 \leqslant b)$ such that $W(t_0)=0$. The linear homogeneous algebraic system

$$c_1\boldsymbol{x}_1(t_0)+c_2\boldsymbol{x}_2(t_0)+\cdots+c_n\boldsymbol{x}_n(t_0)=\boldsymbol{0} \tag{6.2.4}$$

has coefficient determinant $W(t_0)$. As $W(t_0)=0$, (6.2.4) has non-zero solutions $\tilde{c}_1, \tilde{c}_2, \cdots, \tilde{c}_n$. To construct the vector function

$$\boldsymbol{x}(t)=\tilde{c}_1\boldsymbol{x}_1(t)+\tilde{c}_2\boldsymbol{x}_2(t)+\cdots+\tilde{c}_n\boldsymbol{x}_n(t) \tag{6.2.5}$$

because of Superposition Principle, $\boldsymbol{x}(t)$ is the solution of (6.2.2) with initial condition $\boldsymbol{x}(t_0)=\boldsymbol{0}$. Existence and Uniqueness shows that $\boldsymbol{x}(t)=\boldsymbol{0}$. In detail,

$$\tilde{c}_1\boldsymbol{x}_1(t)+\tilde{c}_2\boldsymbol{x}_2(t)+\cdots+\tilde{c}_n\boldsymbol{x}_n(t)=\boldsymbol{0},\ t \in [a,b]$$

Thus, solutions $\boldsymbol{x}_1(t), \boldsymbol{x}_2(t), \cdots, \boldsymbol{x}_n(t)$ are linearly dependent for $t \in [a,b]$. This contradicts the assumption.

From Theorems 6.2.2 and 6.2.3, Wronskian determinant $W(t)$ created by solutions $\boldsymbol{x}_1(t), \boldsymbol{x}_2(t), \cdots, \boldsymbol{x}_n(t)$ to (6.2.2) either equals to zero or never to be zero.

Theorem 6.2.4. There must be n linear independent solutions $\boldsymbol{x}_1(t), \boldsymbol{x}_2(t), \cdots, \boldsymbol{x}_n(t)$ for linear homogeneous system of differential equations (6.2.2).

Proof. For any $t_0 \in [a, b]$, according to Existence and Uniqueness Theorem, there must be solutions $\boldsymbol{x}_1(t), \boldsymbol{x}_2(t), \cdots, \boldsymbol{x}_n(t)$ to (6.2.2) with initial values

$$\boldsymbol{x}_1(t_0)=\begin{bmatrix}1\\0\\0\\\vdots\\0\end{bmatrix}, \boldsymbol{x}_2(t_0)=\begin{bmatrix}0\\1\\0\\\vdots\\0\end{bmatrix}, \cdots, \boldsymbol{x}_n(t_0)=\begin{bmatrix}0\\0\\0\\\vdots\\1\end{bmatrix}$$

Because of Wronskian determinant $W(t_0)=1\neq 0$ of $\boldsymbol{x}_1(t)$, $\boldsymbol{x}_2(t)$, $\cdots$, $\boldsymbol{x}_n(t)$ together with Theorem 6.2.2, $\boldsymbol{x}_1(t), \boldsymbol{x}_2(t), \cdots, \boldsymbol{x}_n(t)$ are linearly independent.

Theorem 6.2.5. If $\boldsymbol{x}_1(t)$, $\boldsymbol{x}_2(t)$, $\cdots$, $\boldsymbol{x}_n(t)$ are n linearly independent solutions to (6.2.2), then any solution $\boldsymbol{x}(t)$ of (6.2.2) can be expressed as

$$\boldsymbol{x}(t)=c_1\boldsymbol{x}_1(t)+c_2\boldsymbol{x}_2(t)+\cdots+c_n\boldsymbol{x}_n(t)$$

where $c_1, c_2, \cdots, c_n$ are corresponding determined constants.

Proof. For any $t_0 \in [a, b]$,

$$\boldsymbol{x}(t_0)=c_1\boldsymbol{x}_1(t_0)+c_2\boldsymbol{x}_2(t_0)+\cdots+c_n\boldsymbol{x}_n(t_0) \tag{6.2.6}$$

is regarded as linear differential equations about $c_1, c_2, \cdots, c_n$, whose coefficient determinant is $W(t_0)$. $\boldsymbol{x}_1(t), \boldsymbol{x}_2(t), \cdots, \boldsymbol{x}_n(t)$ are linearly independent, therefore $W(t_0)\neq 0$ form Theorem 6.2.2. The system of equations (6.2.6) has a unique solution $c_1, c_2, \cdots, c_n$, then the vector function $c_1\boldsymbol{x}_1(t)+c_2\boldsymbol{x}_2(t)+\cdots+c_n\boldsymbol{x}_n(t)$ is a solution of (6.2.2). Thus, we have two solutions $\boldsymbol{x}(t)$ and $c_1\boldsymbol{x}_1(t)+c_2\boldsymbol{x}_2(t)+\cdots+c_n\boldsymbol{x}_n(t)$ with the same initial values to (6.2.2). Uniqueness of solution presents

$$\boldsymbol{x}(t)=c_1\boldsymbol{x}_1(t)+c_2\boldsymbol{x}_2(t)+\cdots+c_n\boldsymbol{x}_n(t)$$

Corollary 6.2.1. The number of linear independent solutions to (6.2.2) is at most n.

Corollary 6.2.2. (6.2.2) can be reduced to a system of linear differential equations with $n-k$ unknown functions based on k known linearly independent solutions. In particular, the general solution of (6.2.2) is obtained from $n-1$ linearly independent solutions.

We call n linearly independent solutions $\boldsymbol{x}_1(t), \boldsymbol{x}_2(t), \cdots, \boldsymbol{x}_n(t)$ fundamental system of solutions[1] of (6.2.2). Obviously, (6.2.2) has infinitely many different fundamental systems of solutions.

From Theorems 6.2.4 and 6.2.5, we know that the solution space of (6.2.2)

---

[1] fundamental system of solutions 基础解系

is n-dimensional, thus the question above has been answered. The main results of this part can be simply expressed: The set of all solutions to (6.2.2) constitutes an n-dimensional linear space.

Now, we write the theorems of this section in the form of matrices. This different expression will be useful in the future. If each column of a $n\times n$ matrix is a solution to (6.2.2), then this matrix is its solution matrix❶. Especially, the solution matrix of (6.2.2) owning linearly independent columns on $[a,b]$ is named fundamental solution matrix❷ written as $\boldsymbol{\Phi}(t)$ with columns $\boldsymbol{\varphi}_1(t), \boldsymbol{\varphi}_2(t), \cdots, \boldsymbol{\varphi}_n(t)$, which is standard fundamental solution matrix❸ when $\boldsymbol{\Phi}(t)\boldsymbol{\Phi}(t_0)=\boldsymbol{E}$ ($\boldsymbol{E}$ is an identity matrix). As a result, Theorems 6.2.4 and 6.2.5 can be expressed as the following theorem.

Theorem 6.2.6. There must be a fundamental solution matrix $\boldsymbol{\Phi}(t)$ to (6.2.2). If $\boldsymbol{\psi}(t)$ is an arbitrary solution of (6.2.2), then

$$\boldsymbol{\psi}(t)=\boldsymbol{\Phi}(t)\boldsymbol{c} \tag{6.2.7}$$

where $\boldsymbol{c}$ is a determined n-dimensional constant column vector.

From the discussion above, in order to find any solution to (6.2.2), we need a fundamental solution matrix. If a solution matrix of (6.2.2) on $[a,b]$ is found, can we verify whether it is a standard solution matrix in some simple way? Theorems 6.2.2 and 6.2.3 fully answer this question, which can be presented in the following form:

Theorem 6.2.7. A solution matrix $\boldsymbol{\Phi}(t)$ of (6.2.2) is a standard solution matrix if and only if $\det\boldsymbol{\Phi}(t)\neq 0$ $(a\leqslant t\leqslant b)$; moreover, if for some $t_0\in[a,b]$, $\det\boldsymbol{\Phi}(t_0)\neq 0$, then $\det\boldsymbol{\Phi}(t)\neq 0$ $(a\leqslant t\leqslant b)$.

It is to be noted that the column vectors of the matrix whose determinant is always equal to zero are not necessarily linearly dependent. For example, the determinant of the matrix

$$\begin{bmatrix} 1 & t & t^2 \\ 0 & 1 & t \\ 0 & 0 & 0 \end{bmatrix}$$

is always zero over any interval, but its column vectors are linearly independent. It is known from Theorem 6.2.7 that this matrix cannot be the solution matrix of any linear homogeneous differential equations.

---

❶ solution matrix 解矩阵

❷ fundamental solution matrix 基解矩阵

❸ standard fundamental solution matrix 标准基解矩阵

【**Example 6.2.3**】 Show that

$$\boldsymbol{\Phi}(t)=\begin{bmatrix} e^t & te^t \\ 0 & e^t \end{bmatrix}$$

is a fundamental solution matrix of system

$$\boldsymbol{x}'=\begin{bmatrix} 1 & 1 \\ 0 & 1 \end{bmatrix}\boldsymbol{x}$$

Solution. Firstly, it is necessary to prove $\boldsymbol{\Phi}(t)$ is a solution matrix. Denote $\boldsymbol{\varphi}_1(t)$ to be the first column of $\boldsymbol{\Phi}(t)$, then

$$\boldsymbol{\varphi}_1'(t)=\begin{bmatrix} e^t \\ 0 \end{bmatrix}=\begin{bmatrix} 1 & 1 \\ 0 & 1 \end{bmatrix}\begin{bmatrix} e^t \\ 0 \end{bmatrix}=\begin{bmatrix} 1 & 1 \\ 0 & 1 \end{bmatrix}\boldsymbol{\varphi}_1(t)$$

meaning $\boldsymbol{\varphi}_1(t)$ is a solution. In the same way, the second column $\boldsymbol{\varphi}_2(t)$ of $\boldsymbol{\Phi}(t)$ is a solution, too. Thus, $\boldsymbol{\Phi}(t)=[\boldsymbol{\varphi}_1(t)\ \boldsymbol{\varphi}_2(t)]$ is a solution matrix.

In addition, according to Theorem 6.2.7, $\det\boldsymbol{\Phi}(t)=e^{2t}\neq 0$ demonstrates that $\boldsymbol{\Phi}(t)$ is a fundamental solution matrix.

Based on Theorems 6.2.6 and 6.2.7, the following corollary is got.

Corollary 6.2.3. Suppose that $\boldsymbol{\Phi}(t)$ $(t\in[a,b])$ is a fundamental solution matrix of (6.2.2), and $\boldsymbol{C}$ is an $n\times n$ nonsingular constant matrix, then $\boldsymbol{\Phi}(t)\boldsymbol{C}$ $(t\in[a,b])$ is also a fundamental solution matrix of (6.2.2).

Proof. Any solution matrix $\boldsymbol{X}(t)$ to (6.2.2) satisfies

$$\boldsymbol{X}'(t)=\boldsymbol{A}(t)\boldsymbol{X}(t),\ a\leqslant t\leqslant b$$

and vice verse. Let

$$\boldsymbol{\Psi}(t)=\boldsymbol{\Phi}(t)\boldsymbol{C},\ a\leqslant t\leqslant b$$

differentiating the equation above, then it is received that

$$\boldsymbol{\Psi}'(t)=\boldsymbol{\Phi}'(t)\boldsymbol{C}=\boldsymbol{A}(t)\boldsymbol{\Phi}(t)\boldsymbol{C}=\boldsymbol{A}(t)\boldsymbol{\Psi}(t)$$

that is, $\boldsymbol{\Psi}(t)$ is a solution matrix of (6.2.2). Because $\boldsymbol{C}$ is nonsingular, we have

$$\det\boldsymbol{\Psi}(t)=\det\boldsymbol{\Phi}(t).\det\boldsymbol{C}\neq 0,\ a\leqslant t\leqslant b.$$

As a result, Theorem 6.2.7 gives that $\boldsymbol{\Psi}(t)=\boldsymbol{\Phi}(t)\boldsymbol{C}$ is a fundamental solution to (6.2.2).

Corollary 6.2.4. If $\boldsymbol{\Phi}(t)$, $\boldsymbol{\Psi}(t)$ are two fundamental solution matrices on $[a,b]$ to $\boldsymbol{x}'=\boldsymbol{A}(t)\boldsymbol{x}$, then there is a nonsingular $n\times n$ constant matrix $\boldsymbol{C}$ such that $\boldsymbol{\Psi}(t)=\boldsymbol{\Phi}(t)\boldsymbol{C}$ $(t\in[a,b])$.

Proof. $\boldsymbol{\Phi}(t)$ is a fundamental solution matrix, so its inverse matrix $\boldsymbol{\Phi}^{-1}(t)$ must exist. Let

$$\boldsymbol{\Phi}^{-1}(t)\boldsymbol{\Psi}(t)=\boldsymbol{X}(t),\ a\leqslant t\leqslant b$$

or

$$\boldsymbol{\Psi}(t)=\boldsymbol{\Phi}(t)\boldsymbol{X}(t),\ a\leqslant t\leqslant b$$

and it is easy to know that $\boldsymbol{X}(t)$ is an $n\times n$ differential matrix as well as $\det\boldsymbol{X}(t)\neq 0\ (a\leqslant t\leqslant b)$. So

$$\begin{aligned}\boldsymbol{A}(t)\boldsymbol{\Psi}(t)&=\boldsymbol{\Psi}'(t)=\boldsymbol{\Phi}'(t)\boldsymbol{X}(t)+\boldsymbol{\Phi}(t)\boldsymbol{X}'(t)\\&=\boldsymbol{A}(t)\boldsymbol{\Phi}(t)\boldsymbol{X}(t)+\boldsymbol{\Phi}(t)\boldsymbol{X}'(t)\\&=\boldsymbol{A}(t)\boldsymbol{\Psi}(t)++\boldsymbol{\Phi}(t)\boldsymbol{X}'(t),\ a\leqslant t\leqslant b\end{aligned}$$

leading to $\boldsymbol{\Phi}(t)\boldsymbol{X}'(t)=\boldsymbol{0}$ or $\boldsymbol{X}'(t)=\boldsymbol{0}\,(a\leqslant t\leqslant b)$, i. e. $\boldsymbol{X}(t)$ is a constant matrix written as $\boldsymbol{C}$. Therefore,

$$\boldsymbol{\Psi}(t)=\boldsymbol{\Phi}(t)\boldsymbol{C},\ a\leqslant t\leqslant b$$

where $\boldsymbol{C}=\boldsymbol{\Phi}^{-1}(t)\boldsymbol{\Psi}(t)$ is a nonsingular $n\times n$ constant matrix.

6. 2. 2 Linear inhomogeneous systems

This section is mainly about the structure of solutions to linear inhomogeneous differential system (6. 2. 1). In (6. 2. 1), $\boldsymbol{A}(t)$ and $\boldsymbol{f}(t)$ are an $n\times n$ continuous matrix and n-dimensional continuous column vector on $[a,b]$, respectively.

Easy to check two simple properties of (6. 2. 1).

Property 6. 2. 1. If $\boldsymbol{\varphi}(t)$ is a solution of (6. 2. 1) and $\boldsymbol{\psi}(t)$ is a solution of its corresponding linear homogeneous differential equations (6. 2. 2), then $\boldsymbol{\varphi}(t)+\boldsymbol{\psi}(t)$ is a solution to (6. 2. 1).

Property 6. 2. 2. If $\tilde{\boldsymbol{\varphi}}(t)$ and $\overline{\boldsymbol{\varphi}}(t)$ are two solutions of (6. 2. 1), then $\tilde{\boldsymbol{\varphi}}(t)-\overline{\boldsymbol{\varphi}}(t)$ is a solution to (6. 2. 2).

The structure of solutions to (6. 2. 1) is declared utilizing the following theorem.

Theorem 6. 2. 8. Assume that $\boldsymbol{\Phi}(t)$ is a fundamental solution matrix of (6. 2. 2), and $\overline{\boldsymbol{\varphi}}$ is some solution of (6. 2. 1), then any solution $\boldsymbol{\varphi}(t)$ of (6. 2. 1) can be presented by

$$\boldsymbol{\varphi}(t)=\boldsymbol{\Phi}(t)\boldsymbol{c}+\overline{\boldsymbol{\varphi}}(t) \tag{6. 2. 8}$$

where $\boldsymbol{c}$ *is a fixed constant column vector.*

Proof. Property 6. 2. 2 shows that $\boldsymbol{\varphi}(t)-\overline{\boldsymbol{\varphi}}(t)$ is a solution to (6. 2. 2). According to Theorem 6. 2. 6, we have

$$\boldsymbol{\varphi}(t)-\overline{\boldsymbol{\varphi}}(t)=\boldsymbol{\Phi}(t)\boldsymbol{c}$$

where $\boldsymbol{c}$ is a fixed constant column vector. Thus,

$$\boldsymbol{\varphi}(t)=\boldsymbol{\Phi}(t)\boldsymbol{c}+\overline{\boldsymbol{\varphi}}(t)$$

Theorem 6. 2. 8 tells us that in order to find any solution of (6. 2. 1), we only need to know a solution of (6. 2. 1) and a fundamental solution matrix of its corresponding linear homogeneous differential equations (6. 2. 2). Now, we have to further point out that in the case of known fundamental solution matrix $\boldsymbol{\Phi}(t)$ of (6. 2. 2), there is a simple method for finding the solution of (6. 2. 1), which is constant variation method[1].

From the previous section we know that if $\boldsymbol{c}$ is a constant column vector, then $\boldsymbol{\varphi}(t)=\boldsymbol{\Phi}(t)\boldsymbol{c}$ is a solution of (6.2.2), which cannot be a solution of (6.2.1). Therefore, we change $\boldsymbol{c}$ to a vector function of t, and try to find a solution of the form

$$\boldsymbol{\varphi}(t)=\boldsymbol{\Phi}(t)\boldsymbol{c}(t) \tag{6.2.9}$$

of (6.2.1), where $\boldsymbol{c}(t)$ is a pending vector function.

Assume that there is a solution to (6.2.1) of the form (6.2.9). At this time, plugging (6.2.9) into (6.2.1),

$$\boldsymbol{\Phi}'(t)\boldsymbol{c}(t)+\boldsymbol{\Phi}(t)\boldsymbol{c}'(t)=\boldsymbol{A}(t)\boldsymbol{\Phi}(t)\boldsymbol{c}(t)+\boldsymbol{f}(t)$$

Because $\boldsymbol{\Phi}(t)$ is a fundamental solution matrix of (6.2.2), $\boldsymbol{\Phi}'(t)=\boldsymbol{A}(t)\boldsymbol{\Phi}(t)$ and the term $\boldsymbol{A}(t)\boldsymbol{\Phi}(t)\boldsymbol{c}(t)$ disappears. $\boldsymbol{c}(t)$ should meet

$$\boldsymbol{\Phi}(t)\boldsymbol{c}'(t)=\boldsymbol{f}(t) \tag{6.2.10}$$

As the nonsingularity of $\boldsymbol{\Phi}(t)$ $(t\in[a,b])$, $\boldsymbol{\Phi}^{-1}(t)$ exists. Left multiplying $\boldsymbol{\Phi}^{-1}(t)$ on both sides of (6.2.10), then integrating it to get

$$\boldsymbol{c}(t)=\int_{t_0}^{t}\boldsymbol{\Phi}^{-1}(s)\boldsymbol{f}(s)\,\mathrm{d}s,\ t_0,\ t\in[a,b]$$

where $\boldsymbol{c}(t_0)=\boldsymbol{0}$. Thus, (6.2.9) is transferred into

$$\boldsymbol{\varphi}(t)=\boldsymbol{\Phi}(t)\int_{t_0}^{t}\boldsymbol{\Phi}^{-1}(s)\boldsymbol{f}(s)\,\mathrm{d}s,\ t_0,\ t\in[a,b] \tag{6.2.11}$$

As a result, if (6.2.1) has a solution $\boldsymbol{\varphi}(t)$ of the form (6.2.9), then $\boldsymbol{\varphi}(t)$ is determined by the formula (6.2.11).

Conversely, the vector function $\boldsymbol{\varphi}(t)$ determined by equation (6.2.11) must be the solution of (6.2.1). In fact, differentiating (6.2.11) gets

$$\begin{aligned}\boldsymbol{\varphi}'(t)&=\boldsymbol{\Phi}'(t)\int_{t_0}^{t}\boldsymbol{\Phi}^{-1}(s)\boldsymbol{f}(s)\,\mathrm{d}s+\boldsymbol{\Phi}(t)\boldsymbol{\Phi}^{-1}(t)\boldsymbol{f}(t)\\&=\boldsymbol{A}(t)\boldsymbol{\Phi}(t)\int_{t_0}^{t}\boldsymbol{\Phi}^{-1}(s)\boldsymbol{f}(s)\,\mathrm{d}s+\boldsymbol{f}(t)\end{aligned}$$

and combining (6.2.11) gives

$$\boldsymbol{\varphi}'(t)=\boldsymbol{A}(t)\boldsymbol{\varphi}(t)+\boldsymbol{f}(t)$$

In this way, we have the following theorem.

Theorem 6.2.9. If $\boldsymbol{\Phi}(t)$ is a fundamental solution matrix of (6.2.2), then the vector function

$$\boldsymbol{\varphi}(t)=\boldsymbol{\Phi}(t)\int_{t_0}^{t}\boldsymbol{\Phi}^{-1}(s)\boldsymbol{f}(s)\,\mathrm{d}s$$

is a solution of (6.2.1) satisfying the initial value $\boldsymbol{\varphi}(t_0)=\boldsymbol{0}$.

---

[1] constant variation method 常数变易法

It is easy to get from Theorems 6.2.8 as well as 6.2.9, that the solution $\boldsymbol{\varphi}(t)$ of (6.2.1) with the initial value condition $\boldsymbol{\varphi}(t_0)=\boldsymbol{\eta}$ is given by the following formula:

$$\boldsymbol{\varphi}(t)=\boldsymbol{\Phi}(t)\boldsymbol{\Phi}^{-1}(t_0)\boldsymbol{\eta}+\boldsymbol{\Phi}(t)\int_{t_0}^{t}\boldsymbol{\Phi}^{-1}(s)\boldsymbol{f}(s)\mathrm{d}s \tag{6.2.12}$$

where $\boldsymbol{\varphi}_h(t)=\boldsymbol{\Phi}(t)\boldsymbol{\Phi}^{-1}(t_0)\boldsymbol{\eta}$ is the solution of (6.2.2) meeting the initial value $\boldsymbol{\varphi}_{\mathrm{h}}(t_0)=\boldsymbol{\eta}$. (6.2.11) or (6.2.12) becomes the constant variation formula[1] of the linear inhomogeneous differential equations (6.2.1).

【Example 6.2.4】 Try to find the solution of the initial value problem

$$\boldsymbol{x}'=\begin{bmatrix}1 & 1\\ 0 & 1\end{bmatrix}\boldsymbol{x}+\begin{bmatrix}\mathrm{e}^{-t}\\ 0\end{bmatrix},\ \boldsymbol{x}(0)=\begin{bmatrix}-1\\ 1\end{bmatrix}$$

Solution. Based on Example 6.2.3, we have known that

$$\boldsymbol{\Phi}(t)=\begin{bmatrix}\mathrm{e}^t & t\mathrm{e}^t\\ 0 & e^t\end{bmatrix}$$

is a fundamental solution matrix of corresponding linear homogeneous differential system. Taking the inverse of the matrix $\boldsymbol{\Phi}(t)$, it follows that

$$\boldsymbol{\Phi}^{-1}(s)=\frac{\begin{bmatrix}\mathrm{e}^s & -s\mathrm{e}^s\\ 0 & \mathrm{e}^s\end{bmatrix}}{\mathrm{e}^{2s}}=\begin{bmatrix}1 & -s\\ 0 & 1\end{bmatrix}\mathrm{e}^{-s}$$

According to Theorem 6.2.9, the solution meeting the initial value $\boldsymbol{\psi}(0)=\begin{bmatrix}0\\ 0\end{bmatrix}$ is

$$\begin{aligned}\boldsymbol{\psi}(t)&=\begin{bmatrix}\mathrm{e}^t & t\mathrm{e}^t\\ 0 & \mathrm{e}^t\end{bmatrix}\int_0^1\mathrm{e}^{-s}\begin{bmatrix}1 & -s\\ 0 & 1\end{bmatrix}\begin{bmatrix}\mathrm{e}^{-s}\\ 0\end{bmatrix}\mathrm{d}s=\begin{bmatrix}\mathrm{e}^t & t\mathrm{e}^t\\ 0 & \mathrm{e}^t\end{bmatrix}\int_0^1\mathrm{e}^{-s}\begin{bmatrix}\mathrm{e}^{-2s}\\ 0\end{bmatrix}\mathrm{d}s\\ &=\begin{bmatrix}\mathrm{e}^t & t\mathrm{e}^t\\ 0 & \mathrm{e}^t\end{bmatrix}\begin{bmatrix}\frac{1}{2}(1-\mathrm{e}^{-2t})\\ 0\end{bmatrix}=\begin{bmatrix}\frac{1}{2}(\mathrm{e}^t-\mathrm{e}^{-t})\\ 0\end{bmatrix}\end{aligned}$$

$\boldsymbol{\Phi}(0)=\boldsymbol{E}$, so the solution with initial value $\boldsymbol{\varphi}_{\mathrm{h}}(0)=\begin{bmatrix}-1\\ 1\end{bmatrix}$ of corresponding linear homogeneous differential system holds

$$\boldsymbol{\varphi}_{\mathrm{h}}(t)=\boldsymbol{\Phi}(t)\begin{bmatrix}-1\\ 1\end{bmatrix}=\begin{bmatrix}(t-1)\mathrm{e}^t\\ \mathrm{e}^t\end{bmatrix}$$

(6.2.12) gives the solution

$$\boldsymbol{\varphi}(t)=\boldsymbol{\varphi}_{\mathrm{h}}(t)+\boldsymbol{\psi}(t)=\begin{bmatrix}(t-1)\mathrm{e}^t\\ \mathrm{e}^t\end{bmatrix}+\begin{bmatrix}\frac{1}{2}(\mathrm{e}^t-\mathrm{e}^{-t})\\ 0\end{bmatrix}=\begin{bmatrix}t\mathrm{e}^t-\frac{1}{2}(\mathrm{e}^t+\mathrm{e}^{-t})\\ \mathrm{e}^t\end{bmatrix}$$

[1] constant variation formula 常数变易公式

6.3 Linear differential systems with constant coefficients

This section studies the problem of linear differential equations with constant coefficients, and mainly discusses the structure of the fundamental solution matrix of linear homogeneous differential equations

$$\boldsymbol{x}'=\boldsymbol{A}\boldsymbol{x} \tag{6.3.1}$$

where $\boldsymbol{A}$ is an $n\times n$ constant matrix. We seek a fundamental solution matrix of (6.3.1) through algebraic methods.

6.3.1 Definition and properties of matrix exponent expA

In order to find a fundamental solution matrix of (6.3.1), we need to define the matrix exponent (or written as $e^{\boldsymbol{A}}$), using the relevant definitions and results of the matrix sequences in Section 6.1.2.

If $\boldsymbol{A}$ is an $n\times n$ constant matrix, then we define the matrix exponent $\exp\boldsymbol{A}$ as the sum of the following matrix series

$$\exp\boldsymbol{A}=\sum_{k=0}^{\infty}\frac{\boldsymbol{A}^k}{k!}=\boldsymbol{E}+\boldsymbol{A}+\frac{\boldsymbol{A}^2}{2!}+\cdots+\frac{\boldsymbol{A}^m}{m!}+\cdots \tag{6.3.2}$$

where $\boldsymbol{E}$ is nth-order identity matrix and $\boldsymbol{A}^m$ is m power of matrix $\boldsymbol{A}$. This series is convergent for all $\boldsymbol{A}$, so $\exp\boldsymbol{A}$ is a certain matrix. Especially, for zero matrix$\boldsymbol{O}$ where all elements are 0, $\exp\boldsymbol{O}=\boldsymbol{E}$.

In fact, it is simple to know for any positive integer k from the properties in Section 6.1.2

$$\left\|\frac{\boldsymbol{A}^k}{k!}\right\|\leqslant\frac{\|\boldsymbol{A}\|^k}{k!}$$

in addition, numerical series

$$\|\boldsymbol{E}\|+\|\boldsymbol{A}\|+\frac{\|\boldsymbol{A}\|^2}{2!}+\cdots+\frac{\|\boldsymbol{A}\|^m}{m!}+\cdots$$

is convergent (its sum is $n-1+e^{\|\boldsymbol{A}\|}$). It is known from Section 6.1.2 that if the norm of each term of a matrix series is less than the corresponding term of a convergent numerical series, then the matrix series is convergent, and thus (6.3.2) is absolutely convergent for any matrix $\boldsymbol{A}$.

It should be further noted that the series

$$\exp\boldsymbol{A}t=\sum_{k=0}^{\infty}\frac{\boldsymbol{A}^k t^k}{k!} \tag{6.3.3}$$

is uniformly convergent over any finite interval of t. Indeed, for any positive integer k, if $|t|\leqslant c$ (c is some positive constant), we have

$$\left\|\frac{\boldsymbol{A}^k t^k}{k!}\right\|\leqslant\frac{\|\boldsymbol{A}\|^k |t|^k}{k!}\leqslant\frac{\|\boldsymbol{A}\|^k c^k}{k!}$$

together with the convergence of numerical series $\sum\limits_{k=0}^{\infty}\frac{(\|\boldsymbol{A}\|c)^k}{k!}$, therefore, (6.3.3) is uniformly convergent. The properties of matrix exponent $\exp\boldsymbol{A}$ are as follows:

(1) If matrices $\boldsymbol{A}$, $\boldsymbol{B}$ are exchangeable ($\boldsymbol{AB}=\boldsymbol{BA}$), then

$$\exp(\boldsymbol{A}+\boldsymbol{B})=\exp\boldsymbol{A}+\exp\boldsymbol{B} \tag{6.3.4}$$

Actually, since the matrix series (6.3.2) is absolutely convergent, some theorems about the operation among absolutely convergent numerical series, for instance, the rearrangement of terms does not change the convergence and the sum of the series and the multiplication theorem of the series, can be used in the matrix series. Binomial theorem and $\boldsymbol{AB}=\boldsymbol{BA}$ show that

$$\exp(\boldsymbol{A}+\boldsymbol{B})=\sum_{k=0}^{\infty}\frac{(\boldsymbol{A}+\boldsymbol{B})^k}{k!}=\sum_{k=0}^{\infty}\left[\sum_{l=0}^{k}\frac{\boldsymbol{A}^l\boldsymbol{B}^{k-l}}{l!\,(k-l)!}\right] \tag{6.3.5}$$

On the other hand, the multiplication theorem of absolutely convergent series gives

$$\exp\boldsymbol{AB}=\sum_{i=0}^{\infty}\frac{\boldsymbol{A}^i}{i!}\sum_{j=0}^{\infty}\frac{\boldsymbol{B}^j}{j!}=\sum_{k=0}^{\infty}\left[\sum_{l=0}^{k}\frac{\boldsymbol{A}^l}{l!}\frac{\boldsymbol{B}^{k-l}}{(k-l)!}\right] \tag{6.3.6}$$

Comparing (6.3.5) and (6.3.6), we get (6.3.4).

(2) For an arbitrary matrix $\boldsymbol{A}$, $(\exp\boldsymbol{A})^{-1}$ exists, and

$$(\exp\boldsymbol{A})^{-1}=\exp(-\boldsymbol{A}) \tag{6.3.7}$$

In reality, $\boldsymbol{A}$ and $-\boldsymbol{A}$ are exchangeable, and let $\boldsymbol{B}-\boldsymbol{A}$ in (6.3.4), thus it is obtained that

$$\exp\boldsymbol{A}\exp(-\boldsymbol{A})=\exp[\boldsymbol{A}+(-\boldsymbol{A})]=\exp\boldsymbol{O}=\boldsymbol{E}$$

that is

$$(\exp\boldsymbol{A})^{-1}=\exp(-\boldsymbol{A})$$

(3) If $\boldsymbol{T}$ is a nonsingular matrix, then

$$\exp(\boldsymbol{T}^{-1}\boldsymbol{AT})=\boldsymbol{T}^{-1}(\exp\boldsymbol{A})\boldsymbol{T}. \tag{6.3.8}$$

In fact

$$\exp(\boldsymbol{T}^{-1}\boldsymbol{AT})=\boldsymbol{E}+\sum_{k=1}^{\infty}\frac{(\boldsymbol{T}^{-1}\boldsymbol{AT})^k}{k!}=\boldsymbol{E}+\sum_{k=1}^{\infty}\frac{\boldsymbol{T}^{-1}\boldsymbol{A}^k\boldsymbol{T}}{k!}=\boldsymbol{E}+\boldsymbol{T}^{-1}\left(\sum_{k=1}^{\infty}\frac{\boldsymbol{A}^k}{k!}\right)\boldsymbol{T}$$
$$=\boldsymbol{T}^{-1}(\exp\boldsymbol{A})\,\boldsymbol{T}$$

Now we can begin to answer the basic questions about linear homogeneous differential equations with constant coefficients (6.3.1).

Theorem 6.3.1. The matrix

$$\boldsymbol{\Phi}(t)=\exp\boldsymbol{A}t \tag{6.3.9}$$

is a fundamental solution matrix to (6.3.1) with $\boldsymbol{\Phi}(0)=\boldsymbol{E}$.

Proof. Easy to know $\boldsymbol{\Phi}(0)=\boldsymbol{E}$. Differentiate t on both sides of (6.3.9) and we reach that

$$\begin{aligned}\boldsymbol{\Phi}'(t)&=(\exp\boldsymbol{A}t)'\\&=\boldsymbol{A}+\frac{\boldsymbol{A}^2t}{1!}+\frac{\boldsymbol{A}^3t^2}{2!}+\cdots+\frac{\boldsymbol{A}^kt^{k-1}}{(k-1)!}+\cdots\\&=\boldsymbol{A}\exp\boldsymbol{A}t=\boldsymbol{A}\boldsymbol{\Phi}(t)\end{aligned}$$

meaning $\boldsymbol{\Phi}(t)$ is a solution matrix of (6.3.1). Because $\det\boldsymbol{\Phi}(0)=\det\boldsymbol{E}=1$, $\boldsymbol{\Phi}(t)$ is a fundamental solution matrix.

By Theorem 6.3.1, we can use this fundamental solution matrix to infer that any solution $\boldsymbol{\varphi}(t)$ of (6.3.1) has the form

$$\boldsymbol{\varphi}(t)=(\exp\boldsymbol{A}t)\boldsymbol{c} \tag{6.3.10}$$

where $\boldsymbol{c}$ is a constant vector.

In some special cases, we can easily get the specific form of the fundamental solution matrix $\exp\boldsymbol{A}t$ of (6.3.1).

【Example 6.3.1】 For the diagonal matrix

$$\boldsymbol{A}=\begin{bmatrix}a_1&&&\\&a_2&&\\&&\ddots&\\&&&a_n\end{bmatrix}$$

find the fundamental solution matrix of $\boldsymbol{x}'=\boldsymbol{A}\boldsymbol{x}$.

Solution. It can be demonstrated from (6.3.2) that

$$\exp\boldsymbol{A}t=\boldsymbol{E}+\begin{bmatrix}a_1&&&\\&a_2&&\\&&\ddots&\\&&&a_n\end{bmatrix}\frac{t}{1!}+\begin{bmatrix}a_1^2&&&\\&a_2^2&&\\&&\ddots&\\&&&a_n^2\end{bmatrix}\frac{t^2}{2!}+\cdots$$

$$+\begin{bmatrix}a_1^k&&&\\&a_2^k&&\\&&\ddots&\\&&&a_n^k\end{bmatrix}\frac{t^k}{k!}+\cdots=\begin{bmatrix}e^{a_1t}&&&\\&e^{a_2t}&&\\&&\ddots&\\&&&e^{a_nt}\end{bmatrix}$$

and according to Theorem 6.3.1, this is a fundamental solution matrix.

【Example 6.3.2】 Find the fundamental solution matrix of $\boldsymbol{x}'=\begin{bmatrix}2&1\\0&2\end{bmatrix}\boldsymbol{x}$.

Solution. Because $\boldsymbol{A}=\begin{bmatrix}2 & 1\\0 & 2\end{bmatrix}=\begin{bmatrix}2 & 0\\0 & 2\end{bmatrix}+\begin{bmatrix}0 & 1\\0 & 0\end{bmatrix}$, which are exchangeable, we have

$$\exp\boldsymbol{A}t=\exp\begin{bmatrix}2 & 0\\0 & 2\end{bmatrix}t\cdot\exp\begin{bmatrix}0 & 1\\0 & 0\end{bmatrix}t$$
$$=\begin{bmatrix}e^{2t} & 0\\0 & e^{2t}\end{bmatrix}\left(\boldsymbol{E}+\begin{bmatrix}0 & 1\\0 & 0\end{bmatrix}t+\begin{bmatrix}0 & 1\\0 & 0\end{bmatrix}^2\frac{t^2}{2!}+\cdots\right)$$

However,

$$\begin{bmatrix}0 & 1\\0 & 0\end{bmatrix}^2=\begin{bmatrix}0 & 0\\0 & 0\end{bmatrix}$$

produces that the series only has two terms. Thus, the fundamental solution matrix is

$$\exp\boldsymbol{A}t=e^{2t}\begin{bmatrix}1 & t\\0 & 1\end{bmatrix}$$

6.3.2 Calculation of fundamental solution matrix

Theorem 6.3.1 tells us that the fundamental solution matrix of (6.3.1) is $\exp\boldsymbol{A}t$, and the problem seems to have been solved. But $\exp\boldsymbol{A}t$ is defined by the matrix series of $\boldsymbol{A}t$, then what is each element of this matrix? In fact, it has not been given except for some specific cases above. This section uses the basic knowledge of linear algebra to discuss the calculation method of $\exp\boldsymbol{A}t$ carefully, so as to solve the structure of the fundamental solution matrix of linear differential equations with constant coefficients.

In order to calculate the fundamental solution matrix $\exp\boldsymbol{A}t$ of (6.3.1), we need to introduce the concept of eigenvalues and eigenvectors of the matrix.

To seek a solution to

$$\boldsymbol{x}'=\boldsymbol{A}\boldsymbol{x} \tag{6.3.1}$$

of the form

$$\boldsymbol{\varphi}(t)=e^{\lambda t}\boldsymbol{c},\boldsymbol{c}\neq\boldsymbol{0} \tag{6.3.11}$$

where the constant λ and the vector $\boldsymbol{c}$ are pending, plug (6.3.11) into (6.3.1) to obtain

$$\lambda e^{\lambda t}\boldsymbol{c}=\boldsymbol{A}e^{\lambda t}\boldsymbol{c}$$

that is,

$$(\lambda\boldsymbol{E}-\boldsymbol{A})\boldsymbol{c}=\boldsymbol{0} \tag{6.3.12}$$

since $e^{\lambda t}\neq 0$, which meaning that $e^{\lambda t}\boldsymbol{c}$ is a solution of (6.3.1) if and only if λ and $\boldsymbol{c}$ meet (6.3.12). (6.3.12) can be regarded as a linear homogeneous algebraic equations of n components of vector $\boldsymbol{c}$. According to the knowledge of linear algebra, this system of equations has a non-zero solution if and only if λ satisfies

$$\det(\lambda\boldsymbol{E}-\boldsymbol{A})=0$$

$\boldsymbol{A}$ being an $n\times n$ constant matrix, the constant λ is an eigenvalue[1] of $\boldsymbol{A}$ if the linear algebraic equations of $\boldsymbol{u}$

$$(\lambda \boldsymbol{E}-\boldsymbol{A})\boldsymbol{u}=\boldsymbol{0} \tag{6.3.13}$$

have a non-zero solution. The non-zero solution $\boldsymbol{u}$ for any eigenvalue λ of (6.3.13) is the eigenvector[2] of $\boldsymbol{A}$ corresponding to the eigenvalue λ.

Nth degree polynomial

$$p(\lambda)=\det(\lambda \boldsymbol{E}-\boldsymbol{A})$$

refers to the characteristic polynomial[3] of $\boldsymbol{A}$, and n-order algebraic equation

$$p(\lambda)=0 \tag{6.3.14}$$

is the characteristic equation of $\boldsymbol{A}$, also called the characteristic equation of (6.3.1).

Thus, under the discussion above, $e^{\lambda t}\boldsymbol{c}$ is the solution of (6.3.1), if and only if λ is the eigenvalue of $\boldsymbol{A}$, and $\boldsymbol{c}$ is the eigenvector corresponding to λ. The eigenvalue of $\boldsymbol{A}$ is the root of the characteristic equation (6.3.14). Since the nth algebraic equation has n roots, $\boldsymbol{A}$ has n eigenvalues, certainly not necessarily different from each other. If $\lambda=\lambda_0$ is a single root (a kth root) of the characteristic equation, then λ_0 is called a simple eigenvalue (a kth eigenvalue)[4].

【Example 6.3.3】 Give the eigenvalues and corresponding eigenvectors of matrix

$$\boldsymbol{A}=\begin{bmatrix} 3 & 5 \\ -5 & 3 \end{bmatrix}$$

Solution. The eigenvalues of $\boldsymbol{A}$ are roots of characteristic equation

$$\det(\lambda \boldsymbol{E}-\boldsymbol{A})=\begin{vmatrix} \lambda-3 & -5 \\ 5 & \lambda-3 \end{vmatrix}=\lambda^2-6\lambda+34=0$$

whose roots are $\lambda_{1,2}=3\pm 5\mathrm{i}$. The eigenvector corresponding to eigenvalue $\lambda_1=3+5\mathrm{i}$

$$\boldsymbol{u}=\begin{bmatrix} u_1 \\ u_2 \end{bmatrix}$$

Satisfies

$$(\lambda_1 \boldsymbol{E}-\boldsymbol{A})\boldsymbol{u}=\begin{bmatrix} 5\mathrm{i} & -5 \\ 5 & 5\mathrm{i} \end{bmatrix}\begin{bmatrix} u_1 \\ u_2 \end{bmatrix}=\boldsymbol{0}$$

that is,

$$\begin{cases} iu_1-u_2=0 \\ u_1-iu_2=0 \end{cases}$$

so for any constant $\alpha\neq 0$,

---

[1] eigenvalue 特征值

[2] eigenvector 特征向量

[3] characteristic polynomial 特征多项式

[4] simple eigenvalue (a kth eigenvalue) 单特征根（k 重特征根）

$$\boldsymbol{u}=\alpha\begin{bmatrix}1\\ \mathrm{i}\end{bmatrix}$$

is the eigenvector corresponding to $\lambda_1=3+5\mathrm{i}$. Similarly, we can get the eigenvector

$$\boldsymbol{v}=\beta\begin{bmatrix}\mathrm{i}\\ 1\end{bmatrix}$$

corresponding to $\lambda_2=3-5\mathrm{i}$ for any constant $\beta\neq 0$.

【Example 6. 3. 4】 Find the eigenvalues and corresponding eigenvectors of matrix

$$\boldsymbol{A}=\begin{bmatrix}2 & 1\\ -1 & 4\end{bmatrix}$$

Solution. The characteristic equation is

$$\det(\lambda\boldsymbol{E}-\boldsymbol{A})=\begin{vmatrix}\lambda-2 & -1\\ 1 & \lambda-4\end{vmatrix}=\lambda^2-6\lambda+9=0$$

so $\lambda=3$ is a double eigenvalue of $\boldsymbol{A}$. Considering the system

$$(3\boldsymbol{E}-\boldsymbol{A})\boldsymbol{c}=\begin{bmatrix}1 & -1\\ 1 & -1\end{bmatrix}\begin{bmatrix}c_1\\ c_2\end{bmatrix}=\boldsymbol{0}$$

or

$$\begin{cases}c_1-c_2=0\\ c_1-c_2=0\end{cases}$$

to get the eigenvector

$$\boldsymbol{c}=\alpha\begin{bmatrix}1\\ 1\end{bmatrix}$$

corresponding to the eigenvalue $\lambda=3$, where $\alpha\neq 0$ is an arbitrary constant.

In Example 6. 3. 3, the eigenvectors $\boldsymbol{u}$ and $\boldsymbol{v}$ are linearly independent because

$$\det[\boldsymbol{u},\boldsymbol{v}]=\begin{vmatrix}\alpha & \beta\mathrm{i}\\ \alpha\mathrm{i} & \beta\end{vmatrix}=2\alpha\beta\neq 0$$

Thus, the vectors $\boldsymbol{u}$, $\boldsymbol{v}$ form the base of the 2-dimensional Euclidean space. However, in Example 6. 3. 4, the eigenvector of $\boldsymbol{A}$ constitutes only a 1-dimensional subspace. It is important to know whether a set of eigenvectors corresponding to individual eigenvalues of a given matrix $\boldsymbol{A}$ constitutes a base. According to the theorem of linear algebra-k eigenvectors corresponding to any k different eigenvalues are linearly independent. Therefore, if the $n\times n$ matrix $\boldsymbol{A}$ has n different eigenvalues, then the corresponding n eigenvectors form a base of the n-dimensional Euclidean space.

First, let us discuss the calculation method of the fundamental solution matrix of differential equations (6. 3. 1) when $\boldsymbol{A}$ has n linearly independent eigenvectors (especially when $\boldsymbol{A}$ has n different eigenvalues).

We can verify the following theorem.

Theorem 6.3.2. If the matrix $\boldsymbol{A}$ has n linearly independent eigenvectors $\boldsymbol{v}_1, \boldsymbol{v}_2, \cdots, \boldsymbol{v}_n$, whose corresponding eigenvalues are $\lambda_1, \lambda_2, \cdots, \lambda_n$ (not necessarily different), then the matrix

$$\boldsymbol{\Phi}(t) = [e^{\lambda_1 t}\boldsymbol{v}_1, e^{\lambda_2 t}\boldsymbol{v}_2, \cdots, e^{\lambda_n t}\boldsymbol{v}_n], -\infty < t < +\infty$$

is a fundamental solution matrix of linear differential equations

$$\boldsymbol{x}' = \boldsymbol{A}\boldsymbol{x}$$

with constant coefficients.

Proof. From the above discussion of eigenvalues and eigenvectors, it is known that each vector function $e^{\lambda_j t}\boldsymbol{v}_j$ $(j=1,2,\cdots,n)$ is a solution of (6.3.1). As a result, the matrix

$$\boldsymbol{\Phi}(t) = [e^{\lambda_1 t}\boldsymbol{v}_1, e^{\lambda_2 t}\boldsymbol{v}_2, \cdots, e^{\lambda_n t}\boldsymbol{v}_n]$$

is a solution matrix of (6.3.1). Again, the linear independence of vectors $\boldsymbol{v}_1, \boldsymbol{v}_2, \cdots, \boldsymbol{v}_n$ leads to

$$\boldsymbol{\Phi}(0) = [\boldsymbol{v}_1, \boldsymbol{v}_2, \cdots, \boldsymbol{v}_n] \neq \boldsymbol{0}$$

Finally, Theorem 6.2.7 in Section 6.2.1 produces that $\boldsymbol{\Phi}(t)$ is a fundamental solution matrix of (6.3.1).

【Example 6.3.5】 Find a fundamental solution matrix of $\boldsymbol{x}' = \boldsymbol{A}\boldsymbol{x}$, where

$$\boldsymbol{A} = \begin{bmatrix} 3 & 5 \\ -5 & 3 \end{bmatrix}$$

Solution. Back to Example 6.3.3, $\lambda_{1,2} = 3 \pm 5i$ are eigenvalues of $\boldsymbol{A}$ and two linearly independent eigenvectors corresponding to $\lambda_{1,2}$ are

$$\boldsymbol{v}_1 = \begin{bmatrix} 1 \\ i \end{bmatrix}, \boldsymbol{v}_2 = \begin{bmatrix} i \\ 1 \end{bmatrix}$$

On the basis of Theorem 6.3.2, the matrix

$$\boldsymbol{\Phi}(t) = \begin{bmatrix} e^{(3+5i)t} & ie^{(3-5i)t} \\ ie^{(3+5i)t} & e^{(3-5i)t} \end{bmatrix}$$

becomes a fundamental solution matrix.

In general, $\boldsymbol{\Phi}(t)$ in Theorem 6.3.2 is not necessarily $\exp\boldsymbol{A}t$. However, according to Corollary 6.2.4 in section 6.2.2, we can determine the relationship between them. Since both $\exp\boldsymbol{A}t$ and $\boldsymbol{\Phi}(t)$ are the fundamental solution matrices of (6.3.1), there is a non-singular constant matrix $\boldsymbol{C}$, making

$$\exp\boldsymbol{A}t = \boldsymbol{\Phi}(t)\boldsymbol{C}$$

In the formula above, $t=0$ leads to $\boldsymbol{C} = \boldsymbol{\Phi}^{-1}(0)$, therefore

$$\exp\boldsymbol{A}t = \boldsymbol{\Phi}(t)\boldsymbol{\Phi}^{-1}(0) \tag{6.3.15}$$

The formula (6.3.15) declares that the calculation of $\exp\boldsymbol{A}t$ is equivalent to that of any of the fundamental solution matrices of the equations (6. 3. 1). Note that equation (6. 3. 15) gives the Remark below.

Remark 6. 3. 1. We know that if $\boldsymbol{A}$ is a real matrix, then $\exp \boldsymbol{A}t$ is the same. Therefore, when $\boldsymbol{A}$ is a real matrix, equation (6. 3. 15) gives a method to construct a real fundamental solution matrix.

【Example 6. 3. 6】 Give a real fundamental solution matrix (or $\exp\boldsymbol{A}t$) of Example (6. 3. 5).

Solution. On the basis of (6. 3. 15) and Remark 6. 3. 1, Example 6. 3. 5 gives that

$$\begin{aligned}\exp\boldsymbol{A}t &= \begin{bmatrix} e^{(3+5i)t} & ie^{(3-5i)t} \\ ie^{(3+5i)t} & e^{(3-5i)t} \end{bmatrix} \begin{bmatrix} 1 & i \\ i & 1 \end{bmatrix}^{-1} \\ &= \frac{1}{2}\begin{bmatrix} e^{(3+5i)t} & ie^{(3-5i)t} \\ ie^{(3+5i)t} & e^{(3-5i)t} \end{bmatrix} \begin{bmatrix} 1 & -i \\ -i & 1 \end{bmatrix} \\ &= \frac{1}{2}\begin{bmatrix} e^{(3+5i)t}+e^{(3-5i)t} & -i(e^{(3+5i)t}-e^{(3-5i)t}) \\ i(e^{(3+5i)t}-e^{(3-5i)t}) & e^{(3+5i)t}+e^{(3-5i)t} \end{bmatrix} \\ &= e^{3t}\begin{bmatrix} \cos 5t & \sin 5t \\ -\sin 5t & \cos 5t \end{bmatrix}\end{aligned}$$

Next, we tend to discuss the calculation of the fundamental solution matrix to (6. 3. 1) for any $n \times n$ matrix $\boldsymbol{A}$. Initially, let us introduce some related knowledge about linear algebra.

$\boldsymbol{A}$ is an $n \times n$ matrix, and $\lambda_1, \lambda_2, \cdots, \lambda_k$ are different eigenvalues of $\boldsymbol{A}$ with multiplicity $n_1, n_2, \cdots, n_k$ $(n_1+n_2+\cdots+n_k=n)$. For each n_jth eigenvalue λ_j, all the solutions of

$$(\boldsymbol{A}-\lambda_j \boldsymbol{E})^{n_j}\boldsymbol{u}=\boldsymbol{0} \qquad (6. 3. 16)$$

constitute a n_j dimensional subspace U_j $(j=1,2,\cdots,k)$ of n-dimensional Euclidean space, which is a direct sum of $U_1, U_2, \cdots, U_k$.

That is to say, for each vector $\boldsymbol{u}$ in n-dimensional Euclidean space, there is unique vectors $\boldsymbol{u}_1, \boldsymbol{u}_2, \cdots, \boldsymbol{u}_k$ with $u_j \in U_j$ $(j=1, 2, \cdots, k)$, so that

$$\boldsymbol{u}=\boldsymbol{u}_1+\boldsymbol{u}_2+\cdots+\boldsymbol{u}_k \qquad (6. 3. 17)$$

Regarding the decomposition formula (6. 3. 17), we cite two special cases. If all the eigenvalues of $\boldsymbol{A}$ are different, that is to say, each $n_j=1$ $(j=1,2,\cdots,k)$, and $k=n$, then for any vector $\boldsymbol{u}$, $\boldsymbol{u}_j$ in (6. 3. 17) can be expressed as $\boldsymbol{u}_j=c_j\boldsymbol{v}_j$, where $\boldsymbol{v}_1, \boldsymbol{v}_2, \cdots, \boldsymbol{v}_n$ are linearly independent eigenvectors of $\boldsymbol{A}$, as well as c_j $(j=1,2,\cdots,n)$ are some constants. If $\boldsymbol{A}$ has only an eigenvalue, i. e. $k=1$, then it is not necessary to decompose the Euclidean space.

Now look at the fundamental solution matrix of (6. 3. 1) using the linear algebra knowledge just quoted. Let's start by looking for any solution $\boldsymbol{\varphi}(t)$ satisfying the initial condition $\boldsymbol{\varphi}(0)=\boldsymbol{\eta}$. From Theorem 6.3.1, we know that $\boldsymbol{\varphi}(t)=(\exp\boldsymbol{A}t)\boldsymbol{\eta}$, and our goal is to calculate $(\exp\boldsymbol{A}t)\boldsymbol{\eta}$ clearly, that is, to know exactly every component of $\boldsymbol{\varphi}(t)$. According to the definition of $\exp\boldsymbol{A}t$, in general, the component of $(\exp\boldsymbol{A}t)\boldsymbol{\eta}$ is an infinite series, so it is difficult to calculate. The key point is to decomposed the initial vector $\boldsymbol{\eta}$ so that the component of $(\exp\boldsymbol{A}t)\boldsymbol{\eta}$ can be written as a linear combination of the finite term of the product between the exponential and the power functions of t.

$\lambda_i\ (i=1,2,\cdots,k)$ is the n_i th distinguished eigenvalue of the matrix $\boldsymbol{A}$.

$$\boldsymbol{\eta}=\boldsymbol{v}_1+\boldsymbol{v}_2+\cdots+\boldsymbol{v}_k \tag{6.3.18}$$

by (6.3.17), where $\boldsymbol{v}_j\in U_j\ (j=1,2,\cdots,k)$. $\boldsymbol{v}_j$ must be the solution of (6.3.16) since the subspace U_j is produced by (6.3.16). We obtain

$$(\boldsymbol{A}-\lambda_j\boldsymbol{E})^l\boldsymbol{v}_j=\boldsymbol{0},\ l\geqslant n_j,\ j=1,2,\cdots,k \tag{6.3.19}$$

Note that for a diagonal matrix, it is simple to get $\exp\boldsymbol{A}t$ so that

$$\mathrm{e}^{\lambda_j t}\exp(-\lambda_j\boldsymbol{E}t)=\mathrm{e}^{\lambda_j t}\begin{bmatrix}\mathrm{e}^{-\lambda_j t} & & & \\ & \mathrm{e}^{-\lambda_j t} & & \\ & & \mathrm{e}^{-\lambda_j t} & \\ & & & \mathrm{e}^{-\lambda_j t}\end{bmatrix}=\boldsymbol{E}$$

together with (6.3.19), so we have

$$\begin{aligned}(\exp\boldsymbol{A}t)\boldsymbol{v}_j&=(\exp\boldsymbol{A}t)\mathrm{e}^{\lambda_j t}\left[\exp(-\lambda_j\boldsymbol{E}t)\right]\boldsymbol{v}_j\\&=\mathrm{e}^{\lambda_j t}\left[\exp(\boldsymbol{A}-\lambda_j\boldsymbol{E}t)\right]\boldsymbol{v}_j\\&=\mathrm{e}^{\lambda_j t}\left[\boldsymbol{E}+t(\boldsymbol{A}-\lambda_j\boldsymbol{E})+\frac{t^2}{2!}(\boldsymbol{A}-\lambda_j\boldsymbol{E})^2+\cdots+\frac{t^{n_j-1}}{(n_j-1)!}(\boldsymbol{A}-\lambda_j\boldsymbol{E})^{n_j-1}\right]\boldsymbol{v}_j\end{aligned}$$

Again, (6.3.18) shows that the solution $\boldsymbol{\varphi}(t)=(\exp\boldsymbol{A}t)\boldsymbol{\eta}$ of (6.3.1) is transformed into

$$\begin{aligned}\boldsymbol{\varphi}(t)&=(\exp\boldsymbol{A}t)\boldsymbol{\eta}=\boldsymbol{\varphi}(t)=(\exp\boldsymbol{A}t)\sum_{j=1}^{k}\boldsymbol{v}_j=\sum_{j=1}^{k}(\exp\boldsymbol{A}t)\boldsymbol{v}_j\\&=\sum_{j=1}^{k}\mathrm{e}^{\lambda_j t}\left[\boldsymbol{E}+t(\boldsymbol{A}-\lambda_j\boldsymbol{E})+\frac{t^2}{2!}(\boldsymbol{A}-\lambda_j\boldsymbol{E})^2+\cdots+\frac{t^{n_j-1}}{(n_j-1)!}(\boldsymbol{A}-\lambda_j\boldsymbol{E})^{n_j-1}\right]\boldsymbol{v}_j\end{aligned}$$

finally the solution $\boldsymbol{\varphi}(t)$ of (6.3.1) with $\boldsymbol{\varphi}(0)=\boldsymbol{\eta}$ can be written as

$$\boldsymbol{\varphi}(t)=\sum_{j=1}^{k}\mathrm{e}^{\lambda_j t}\left[\sum_{i=0}^{n_j-1}\frac{\mathrm{e}^i}{i!}(\boldsymbol{A}-\lambda_j\boldsymbol{E})^i\boldsymbol{v}_j\right] \tag{6.3.20}$$

Especially, if $\boldsymbol{A}$ has only one eigenvalue, then it is not necessary to decompose the initial vector into (6.3.18).

For any $\boldsymbol{u}$, it follows that

$$(\boldsymbol{A}-\lambda \boldsymbol{E})^n \boldsymbol{u}=\boldsymbol{0}$$

that is, $(\boldsymbol{A}-\lambda \boldsymbol{E})^n$ is a zero matrix, and then referring to the definition of $\exp \boldsymbol{A}t$

$$\begin{aligned}\exp \boldsymbol{A}t &= e^{\lambda t} \exp(\boldsymbol{A}-\lambda \boldsymbol{E})t \\ &= e^{\lambda t} \sum_{i=0}^{n-1} \frac{t^i}{i!}(\boldsymbol{A}-\lambda \boldsymbol{E})^i \end{aligned} \tag{6.3.21}$$

In order to get $\exp \boldsymbol{A}t$ from (6.3.20), observing

$$\exp \boldsymbol{A}t=(\exp \boldsymbol{A}t)\boldsymbol{E}=[(\exp \boldsymbol{A}t)\boldsymbol{e}_1,\ (\exp \boldsymbol{A}t)\boldsymbol{e}_2,\cdots,(\exp \boldsymbol{A}t)\boldsymbol{e}_n]$$

where

$$\boldsymbol{e}_1=\begin{bmatrix}1\\0\\\vdots\\0\\0\end{bmatrix},\boldsymbol{e}_2=\begin{bmatrix}0\\1\\\vdots\\0\\0\end{bmatrix},\cdots,\boldsymbol{e}_n=\begin{bmatrix}0\\0\\\vdots\\0\\1\end{bmatrix}$$

are identity matrices, so find n solutions of (6.3.1) by substituting $\boldsymbol{\eta}=\boldsymbol{e}_1$, $\boldsymbol{\eta}=\boldsymbol{e}_2,\cdots,$ $\boldsymbol{\eta}=\boldsymbol{e}_n$ and put them as columns to get $\exp \boldsymbol{A}t$.

【Example 6.3.7】 Solve the initial value problem $\boldsymbol{x}'=\boldsymbol{A}\boldsymbol{x}$, $\boldsymbol{\varphi}(0)=\boldsymbol{\eta}$ and $\exp \boldsymbol{A}t$, where $\boldsymbol{A}$ is the matrix in Example 6.3.4.

Solution. Based on Example 6.3.4, $\lambda_1=3$ is a double eigenvalue of $\boldsymbol{A}$, meaning $n_1=2$ and only one subspace U_1, then plugging $n_1=2$ and $\boldsymbol{\eta}=\begin{bmatrix}\eta_1\\\eta_2\end{bmatrix}$ into (6.3.20) results in

$$\begin{aligned}\boldsymbol{\varphi}(t) &= e^{3t}[\boldsymbol{E}+t(\boldsymbol{A}-3\boldsymbol{E})]\boldsymbol{\eta} \\ &= e^{3t}\left(\boldsymbol{E}+t\begin{bmatrix}-1 & 1\\-1 & 1\end{bmatrix}\right)\begin{bmatrix}\eta_1\\\eta_2\end{bmatrix}=e^{3t}\begin{bmatrix}\eta_1+t(-\eta_1+\eta_2)\\\eta_2+t(-\eta_1+\eta_2)\end{bmatrix}\end{aligned} \tag{6.3.22}$$

(6.3.21) gives that

$$\begin{aligned}\exp \boldsymbol{A}t &= e^{3t}[\boldsymbol{E}+t(\boldsymbol{A}-3\boldsymbol{E})] \\ &= e^{3t}\left(\begin{bmatrix}1 & 0\\0 & 1\end{bmatrix}+t\begin{bmatrix}-1 & 1\\-1 & 1\end{bmatrix}\right)=\begin{bmatrix}1-t & t\\-t & 1+t\end{bmatrix}\end{aligned}$$

Or putting

$$\boldsymbol{\eta}=\boldsymbol{e}_1=\begin{bmatrix}1\\0\end{bmatrix},\ \boldsymbol{\eta}=\boldsymbol{e}_1=\begin{bmatrix}0\\1\end{bmatrix}$$

into (6.3.22) to get

$$\exp \boldsymbol{A}t=e^{3t}\begin{bmatrix}1-t & t\\-t & 1+t\end{bmatrix}$$

【Example 6.3.8】 Find $\exp \boldsymbol{A}t$ where

$$
\boldsymbol{A}=\begin{bmatrix} -4 & 1 & 0 & 0 & 0 \\ 0 & -4 & 1 & 0 & 0 \\ 0 & 0 & -4 & 0 & 0 \\ 0 & 0 & 0 & -4 & 0 \\ 0 & 0 & 0 & 0 & -4 \end{bmatrix}
$$

Solution. In fact, $n=5$, $\lambda=-4$ is a fifth eigenvalue of $\boldsymbol{A}$, and $(\boldsymbol{A}+4\boldsymbol{E})^3=\boldsymbol{0}$ by direct calculation. (6.3.21) gives

$$
\exp \boldsymbol{A}t=\mathrm{e}^{-4t}\left[\boldsymbol{E}+t(\boldsymbol{A}+4\boldsymbol{E})+\frac{t^2}{2!}(\boldsymbol{A}+4\boldsymbol{E})^2\right]
$$

that is,

$$
\exp \boldsymbol{A}t=\mathrm{e}^{-4t}\left(\begin{bmatrix} 1 & & & & \\ & 1 & & & \\ & & 1 & & \\ & & & 1 & \\ & & & & 1 \end{bmatrix}+t\begin{bmatrix} 0 & 1 & 0 & 0 & 0 \\ 0 & 0 & 1 & 0 & 0 \\ 0 & 0 & 0 & 0 & 0 \\ 0 & 0 & 0 & 0 & 0 \\ 0 & 0 & 0 & 0 & 0 \end{bmatrix}+\frac{t^2}{2!}\begin{bmatrix} 0 & 0 & 1 & 0 & 0 \\ 0 & 0 & 0 & 0 & 0 \\ 0 & 0 & 0 & 0 & 0 \\ 0 & 0 & 0 & 0 & 0 \\ 0 & 0 & 0 & 0 & 0 \end{bmatrix}\right)
$$

$$
=\mathrm{e}^{-4t}\begin{bmatrix} 1 & t & \frac{t^2}{2!} & 0 & 0 \\ 0 & 1 & t & 0 & 0 \\ 0 & 0 & 1 & 0 & 0 \\ 0 & 0 & 0 & 1 & 0 \\ 0 & 0 & 0 & 0 & 1 \end{bmatrix}
$$

【Example 6.3.9】 Considering the system

$$
\begin{cases} x'_1=3x_1-x_2+x_3 \\ x'_2=2x_1+x_3 \\ x'_3=x_1-x_2+2x_3 \end{cases}
$$

with the coefficient matrix

$$
\boldsymbol{A}=\begin{bmatrix} 3 & -1 & 1 \\ 2 & 0 & 1 \\ 1 & -1 & 2 \end{bmatrix}
$$

find the solution $\boldsymbol{\varphi}(t)$ meeting the initial value $\boldsymbol{\varphi}(0)=\boldsymbol{\eta}$ and $\exp \boldsymbol{A}t$.

Solution. The characteristic equation of $\boldsymbol{A}$ is

$$
\det(\lambda\boldsymbol{E}-\boldsymbol{A})=(\lambda-1)(\lambda-2)^2=0
$$

with a single root $(n_1=1)$ $\lambda_1=1$ and a double root $(n_2=2)$ $\lambda_2=2$. In order to seek the subspaces U_1 and U_2 in 3-dimensional Euclidean space, (6.3.16) demonstrates that we need to consider the system of equations

$$
(\boldsymbol{A}-\boldsymbol{E})\boldsymbol{u}=\boldsymbol{0} \quad \text{and} \quad (\boldsymbol{A}-2\boldsymbol{E})^2\boldsymbol{u}=\boldsymbol{0}
$$

Initially, discuss

$$(\boldsymbol{A}-\boldsymbol{E})\boldsymbol{u}=\begin{bmatrix}2 & -1 & 1\\ 2 & -1 & 1\\ 1 & -1 & 1\end{bmatrix}\boldsymbol{u}=\boldsymbol{0}$$

or

$$\begin{cases}2u_1-u_2+u_3=0\\ 2u_1-u_2+u_3=0\\ u_1-u_2+u_3=0\end{cases}$$

to obtain the solution $\boldsymbol{u}_1=(0,\alpha,\alpha)^{\mathrm{T}}$, where α is an arbitrary constant. The subspace U_1 is produced by the vector $\boldsymbol{u}_1$.

Further, for

$$(\boldsymbol{A}-2\boldsymbol{E})^2\boldsymbol{u}=\begin{bmatrix}0 & 0 & 0\\ -1 & 1 & 0\\ -1 & 1 & 0\end{bmatrix}\boldsymbol{u}=\boldsymbol{0}$$

or

$$\begin{cases}-u_1+u_2=0\\ -u_1+u_2=0\end{cases}$$

its solution is $\boldsymbol{u}_2=(\beta,\beta,\gamma)^{\mathrm{T}}$, where β, γ are any constants, which produces the subspace U_2.

Now we should find vectors $\boldsymbol{v}_1\in U_1$, $\boldsymbol{v}_2\in U_2$ so that the initial vector $\boldsymbol{\eta}$ is written as the form like (6.3.18). Since $\boldsymbol{v}_1\in U_1$, $\boldsymbol{v}_2\in U_2$,

$$\boldsymbol{v}_1=\begin{bmatrix}0\\ \alpha\\ \alpha\end{bmatrix},\boldsymbol{v}_2=\begin{bmatrix}\beta\\ \beta\\ \gamma\end{bmatrix}$$

where α, β, γ are some constants, that is to say,

$$\begin{bmatrix}\eta_1\\ \eta_2\\ \eta_3\end{bmatrix}=\begin{bmatrix}\beta\\ \alpha+\beta\\ \alpha+\gamma\end{bmatrix}$$

so solving $\beta=\eta_1$, $\alpha+\beta=\eta_2$, $\alpha+\gamma=\eta_3$ gives $\alpha=\eta_2-\eta_1$, $\beta=\eta_1$, $\gamma=\eta_3-\eta_2+\eta_1$, and

$$\boldsymbol{v}_1=\begin{bmatrix}0\\ \eta_2-\eta_1\\ \eta_2-\eta_1\end{bmatrix},\ \boldsymbol{v}_2=\begin{bmatrix}\eta_1\\ \eta_1\\ \eta_3-\eta_2+\eta_1\end{bmatrix}$$

According to (6.3.20), we get the solution with the initial value $\boldsymbol{\varphi}(0)=\boldsymbol{\eta}$

$$\boldsymbol{\varphi}(t)=\mathrm{e}^t\boldsymbol{E}\boldsymbol{v}_1+\mathrm{e}^{2t}\,[\boldsymbol{E}+t(\boldsymbol{A}-2\boldsymbol{E})]\boldsymbol{v}_2$$

$$=e^t\begin{bmatrix}0\\ \eta_2-\eta_1\\ \eta_2-\eta_1\end{bmatrix}+e^{2t}\left(\boldsymbol{E}+t\begin{bmatrix}1&-1&1\\2&-2&1\\1&-1&0\end{bmatrix}\right)\begin{bmatrix}\eta_1\\ \eta_1\\ \eta_3-\eta_2+\eta_1\end{bmatrix}$$

$$=e^t\begin{bmatrix}0\\ \eta_2-\eta_1\\ \eta_2-\eta_1\end{bmatrix}+e^{2t}\begin{bmatrix}1+t&-t&t\\2t&1-2t&t\\t&-t&1\end{bmatrix}\begin{bmatrix}\eta_1\\ \eta_1\\ \eta_3-\eta_2+\eta_1\end{bmatrix}$$

$$=e^t\begin{bmatrix}0\\ \eta_2-\eta_1\\ \eta_2-\eta_1\end{bmatrix}+e^{2t}\begin{bmatrix}\eta_1+t(\eta_3-\eta_2+\eta_1)\\ \eta_1+t(\eta_3-\eta_2+\eta_1)\\ \eta_3-\eta_2+\eta_1\end{bmatrix}$$

To write $\exp\boldsymbol{A}t$, set $\boldsymbol{\eta}=(1,0,0)^T$, $(0,1,0)^T$, $(0,0,1)^T$ in the formula above to get three linearly independent solution, which are used as the columns, then we have

$$\exp\boldsymbol{A}t=\begin{bmatrix}(1+t)e^{2t}&-te^{2t}&te^{2t}\\-e^t+(1+t)e^{2t}&e^t-te^{2t}&te^{2t}\\-e^t+e^{2t}&e^t-e^{2t}&e^{2t}\end{bmatrix}$$

It should be noted that the formula (6. 3. 20) is the main result in this section. Equation (6. 3. 20) tells us that any solution to a system of linear differential equations with constant coefficients (6. 3. 1) can be obtained by finite-time algebraic operations. In the theory and application of ordinary differential equations, the study about the behavior of differential equations when $t\to\infty$ is very important. For example, the stability of the solution of the differential equations to be discussed in Chapter 7 is an aspect. As a direct application of the formula (6. 3. 20), we can get the following Theorem 6. 3. 3. A more profound application of the formula (6. 3. 20) is left to be discussed in Chapter 7.

Theorem 6. 3. 3. For a system of linear differential equations with constant coefficients

$$\boldsymbol{x}'=\boldsymbol{A}\boldsymbol{x}$$

(1) If the real parts of all the eigenvalues of $\boldsymbol{A}$ are negative, then any solution of (6. 3. 1) tends to zero as $t\to+\infty$;

(2) If the real parts of all the eigenvalues of $\boldsymbol{A}$ are non-positive and those with zero real parts are single, then any solution of (6. 3. 1) keeps bounded as $t\to+\infty$;

(3) If the real part of at least one eigenvalue of $\boldsymbol{A}$ are positive, then at least one solution of (6. 3. 1) goes to infinity as $t\to+\infty$.

Proof. According to the formula (6. 3. 20), it is known that any solution of the system (6. 3. 1) can be expressed as a linear combination of the exponential and the product of the power function of t, and then based on the simple properties of the exponential function and the assumptions made in (1) as well as (2), you can get the proof about the first two parts. To prove (3), let $\lambda=\alpha+i\beta$ be the eigenvalue of $\boldsymbol{A}$, where α, β are real numbers and $\alpha>0$. Taking $\boldsymbol{\eta}$ is the eigenvector of $\boldsymbol{A}$ corresponding to the eigenvalue λ, then the vector function

$$\boldsymbol{\varphi}(t)=e^{\lambda t}\boldsymbol{\eta}$$

is a solution of (6.3.1), and

$$\|\boldsymbol{\varphi}(t)\|=e^{\alpha t}\|\boldsymbol{\eta}\|\to\infty\quad(t\to\infty)$$

As a result, the proof is completed.

The steps and formula (6.3.20) discussed in this part provide a method for actually calculating the fundamental solution matrix of (6.3.1). Here we mainly apply the conclusions about spatial decomposition. In fact, other methods of calculating the fundamental solution matrix $\exp\boldsymbol{A}t$ can be obtained by the algebra knowledge else. For example, one of them is to produce Jordan canonical form. This method is quite simple in theory, however it can be quite cumbersome to calculate. The method that now be mentioned is described below as a note for the reader's reference.

Remark 6.3.2. Calculate the fundamental solution matrix using Jordan canonical form.

From theories about the matrix, there exists a nonsigular matrix $\boldsymbol{T}$ such that

$$\boldsymbol{T}^{-1}\boldsymbol{A}\boldsymbol{T}=\boldsymbol{J}\tag{6.3.23}$$

where $\boldsymbol{J}$ has Jordan canonical form

$$\boldsymbol{J}=\begin{bmatrix}\boldsymbol{J}_1 & & & \\ & \boldsymbol{J}_2 & & \\ & & \ddots & \\ & & & \boldsymbol{J}_l\end{bmatrix}$$

with n_jth order matrix

$$\boldsymbol{J}_j=\begin{bmatrix}\lambda_j & 1 & & & \\ & \lambda_j & 1 & & \\ & & \ddots & \ddots & \\ & & & \ddots & 1 \\ & & & & \lambda_j\end{bmatrix}\quad(j=1,2,\cdots,l)$$

$n_1+n_2+\cdots+n_l=n$ and l is the number of primary factors to the matrix $\boldsymbol{A}-\lambda\boldsymbol{E}$. $\lambda_1,\lambda_2,\cdots,\lambda_l$ are the roots of the characteristic equation (6.3.14) some of that may be the same.

Based on the special forms of $\boldsymbol{J}$ and $\boldsymbol{J}_j$ $(j=1,2,\cdots,l)$, (6.3.2) brings in

$$\exp \boldsymbol{J}t=\begin{bmatrix} \exp \boldsymbol{J}_1 t & & & \\ & \exp \boldsymbol{J}_2 t & & \\ & & \ddots & \\ & & & \exp \boldsymbol{J}_l t \end{bmatrix} \tag{6.3.24}$$

where

$$\exp \boldsymbol{J}_j t=\begin{bmatrix} 1 & t & \frac{t^2}{2!} & \cdots & \frac{t^{n_j-1}}{(n_j-1)!} \\ & 1 & t & \cdots & \frac{t^{n_j-2}}{(n_j-2)!} \\ & & \ddots & & \vdots \\ & & & \ddots & \vdots \\ & & & & 1 \end{bmatrix} e^{\lambda_j t} \tag{6.3.25}$$

Thus, we can get $\exp \boldsymbol{J}t$ for Jordan canonical form $\boldsymbol{J}$, and from (6.3.23) and the property (3) of the matrix index, the fundamental solution matrix $\exp \boldsymbol{A}t$ of (6.3.1) is shown

$$\exp \boldsymbol{A}t=\exp(\boldsymbol{T}\boldsymbol{J}\boldsymbol{T}^{-1})t=\exp \boldsymbol{T}(\exp \boldsymbol{J}t)\boldsymbol{T}^{-1} \tag{6.3.26}$$

Certainly, Corollary 6.2.3 in Section 6.2.1 demonstrates

$$\boldsymbol{\psi}(t)=\boldsymbol{T}\exp \boldsymbol{J}t \tag{6.3.27}$$

is also the fundamental matrix solution of (6.3.1). We obtain the specific structure of the fundamental solution matrix from (6.3.26) or (6.3.27), but it is difficult to give the nonsigular matrix.

Ultimately, we investigate the variation of constants formula of the system of linear inhomogeneous equations

$$\boldsymbol{x}'=\boldsymbol{A}\boldsymbol{x}+\boldsymbol{f}(t) \tag{6.3.28}$$

where $\boldsymbol{A}$ is an $n \times n$ constant matrix and $\boldsymbol{f}(t)$ is a continuous vector function. Since the fundamental solution matrix of the linear homogeneous differential equations (6.3.1) corresponding to (6.3.28) is $\boldsymbol{\Phi}=\exp \boldsymbol{A}t$, the form of the variation of constants formula in 6.2.2 becomes very simple. Thus, we have $\boldsymbol{\Phi}^{-1}(s)=\exp(-s\boldsymbol{A})$, $\boldsymbol{\Phi}(t)\boldsymbol{\Phi}^{-1}(s)=\exp[\boldsymbol{A}(t-s)]$, and provided that the initial value is $\boldsymbol{\varphi}(t_0)=\boldsymbol{\eta}$, then $\boldsymbol{\varphi}_k(t)=\exp[(t-t_0)\boldsymbol{A}]\boldsymbol{\eta}$, that is to say, the solution of (6.3.28) is

$$\boldsymbol{\varphi}(t)=\exp[(t-t_0)\boldsymbol{A}]\boldsymbol{\eta}+\int_{t_0}^{t}\exp[(t-s)\boldsymbol{A}]\boldsymbol{f}(s)\,ds \tag{6.3.29}$$

We can use the methods provided in this part to construct a fundamental solution matrix $\exp \boldsymbol{A}t$. However, unless a special case, it is not easy to calculate the integral form in (6.3.29).

【Example 6. 3. 10】 Suppose that

$$\boldsymbol{A}=\begin{bmatrix}3 & 5\\ -5 & 3\end{bmatrix},\boldsymbol{f}(t)=\begin{bmatrix}e^{-t}\\ 0\end{bmatrix}$$

Try to give the solution $\boldsymbol{\varphi}(t)$ of $\boldsymbol{x}'=\boldsymbol{A}x+\boldsymbol{f}(t)$ with the initial value $\boldsymbol{\varphi}(0)=\begin{bmatrix}0\\ 1\end{bmatrix}$.

Solution. Due to the above Example 6. 3. 3, we know

$$\exp\boldsymbol{A}t=e^{3t}\begin{bmatrix}\cos5t & \sin5t\\ -\sin5t & \cos5t\end{bmatrix}$$

Substitute it into (6. 3. 29) to get (using $t_0=0$)

$$\boldsymbol{\varphi}(t)=e^{3t}\begin{bmatrix}\cos5t & \sin5t\\ -\sin5t & \cos5t\end{bmatrix}\begin{bmatrix}0\\ 1\end{bmatrix}$$

$$+\int_0^t e^{3(t-s)}\begin{bmatrix}\cos5(t-s) & \sin5(t-s)\\ -\sin5(t-s) & \cos5(t-s)\end{bmatrix}\begin{bmatrix}e^{-s}\\ 0\end{bmatrix}ds$$

We calculate the above integral as follows

$$\boldsymbol{\varphi}(t)=e^{3t}\begin{bmatrix}\sin5t\\ \cos5t\end{bmatrix}+e^{3t}\int_0^t e^{-4s}\begin{bmatrix}\cos5t\cos5s+\sin5t\sin5s\\ -\sin5t\cos5s+\cos5t\sin5s\end{bmatrix}ds$$

and by integration by parts,

$$\int_0^t e^{-4s}\cos5s\,ds=\frac{e^{-4s}}{16+25}(-4\cos5s+5\sin5s)\Bigg|_{s=0}^{s=t}$$

$$\int_0^t e^{-4s}\sin5s\,ds=\frac{e^{-4s}}{16+25}(-4\sin5s-5\cos5s)\Bigg|_{s=0}^{s=t}$$

as a result, the solution is

$$\boldsymbol{\varphi}(t)=\frac{1}{41}e^{3t}\begin{bmatrix}4\cos5t+46\sin5t-4e^{-4t}\\ 46\cos5t-4\sin5t-5e^{-4t}\end{bmatrix}$$

6. 4 Exercises

1. Given the system of differential equations

$$\boldsymbol{x}'=\begin{bmatrix}0 & 1\\ -1 & 0\end{bmatrix}\boldsymbol{x},\boldsymbol{x}=\begin{bmatrix}x_1\\ x_2\end{bmatrix} \qquad (*)$$

(1) Try to certificate that $\boldsymbol{u}(t)=\begin{bmatrix}\cos t\\ -\sin t\end{bmatrix}$ and $\boldsymbol{v}(t)=\begin{bmatrix}\sin t\\ -\cos t\end{bmatrix}$ are solutions to system (*) with initial values $\boldsymbol{u}(0)=\begin{bmatrix}1\\ 0\end{bmatrix}$ and $\boldsymbol{v}(0)=\begin{bmatrix}0\\ 1\end{bmatrix}$, respectively.

(2) Show that $\boldsymbol{w}(t)=c_1\boldsymbol{u}(t)+c_2\boldsymbol{u}(t)$ is the solution to (*) satisfying $\boldsymbol{w}(0)=$

$\begin{bmatrix} c_1 \\ c_2 \end{bmatrix}$, where c_1, c_2 are arbitrary constants.

2. Transform the following problems into the initial value problems of the first-order equations

(1) $x''+2x'+7tx=e^{-t}$, $x(1)=7$, $x'(1)=-2$;

(2) $x^{(4)}+x=te^t$, $x(0)=1$, $x'(0)=-1$, $x''(0)=2$, $x'''(0)=0$;

(3) $\begin{cases} x''+5y'-7x+6y=e^t, \\ y''-2y+13y'-15x=\cos t, \end{cases}$ $x(0)=1$, $x'(0)=0$, $y(0)=0$, $y'(0)=1$.

3. Use the stepwise approximation method to find the third approximate solution of the system of equations

$$\boldsymbol{x}'=\begin{bmatrix} 0 & 1 \\ -1 & 0 \end{bmatrix}\boldsymbol{x}$$

satisfying the initial value

$$\boldsymbol{x}(0)=\begin{bmatrix} 0 \\ 1 \end{bmatrix}.$$

4. Check

$$\boldsymbol{\Phi}(t)=\begin{bmatrix} t^2 & t \\ 2t & 1 \end{bmatrix}$$

to be a fundamental solution matrix of the following system on $[a,b]$ non containing the origin

$$\boldsymbol{x}'=\begin{bmatrix} 0 & 1 \\ -\dfrac{2}{t^2} & \dfrac{2}{t} \end{bmatrix}\boldsymbol{x}$$

5. Consider the differential equations

$$\boldsymbol{x}'=\boldsymbol{A}(t)\boldsymbol{x} \tag{6.2.2}$$

where $\boldsymbol{A}(t)$ is a $n\times n$ continuous matrix on $[a,b]$ with elements $a_{ij}(t)$, $(i,j=1,2,\cdots,n)$.

(1) Suppose that $\boldsymbol{x}_1(t),\boldsymbol{x}_2(t),\cdots,\boldsymbol{x}_n(t)$ are any n solutions to (6.2.2), then their Wronskian determinant $\boldsymbol{W}[\boldsymbol{x}_1(t),\boldsymbol{x}_2(t),\cdots,\boldsymbol{x}_n(t)]=\boldsymbol{W}(t)$ fits a first-order linear differential equation

$$\boldsymbol{W}'=[a_{11}(t)+a_{22}(t)+\cdots+a_{nn}(t)]\boldsymbol{W}$$

(Hint: Use the differential formula of the determinant to find the expression of $\boldsymbol{W}$.)

(2) Find the solutions of the first-order linear differential equation above, and prove the formula:

$$\boldsymbol{W}(t)=\boldsymbol{W}(t_0)e^{\int_{t_0}^{t}[a_{11}(t)+a_{22}(t)+\cdots+a_{nn}(t)]\,ds}, \quad t_0, t\in[a,b]$$

6. $\boldsymbol{A}(t)$ is a continuous $n\times n$ real matrix over the interval $[a,b]$, and $\boldsymbol{\Phi}(t)$, $\boldsymbol{x}=$

$\boldsymbol{\varphi}(t)$ are a fundamental solution matrix and a special solution, especially. Try to prove that:

(1) Any solution $\boldsymbol{y}=\boldsymbol{\psi}(t)$ of $\boldsymbol{y}'=-\boldsymbol{A}^{\mathrm{T}}(t)\boldsymbol{y}$ must satisfy $\boldsymbol{\psi}^{\mathrm{T}}(t)\boldsymbol{\varphi}(t)=$constant.

(2) $\boldsymbol{\Psi}(t)$ is a fundamental solution matrix of $\boldsymbol{y}'=-\boldsymbol{A}^{\mathrm{T}}(t)\boldsymbol{y}$ if and only if there exists nonsingular constant matrix $\boldsymbol{C}$ such that $\boldsymbol{\Psi}^{\mathrm{T}}(t)\boldsymbol{\Phi}(t)=\boldsymbol{C}$.

7. $\boldsymbol{x}(t)=(\varphi_1(t),\varphi_2(t),\cdots,\varphi_n(t))^{\mathrm{T}}$ is a nonzero solution of (6.2.2) with $\varphi_n(t)\neq 0$, and prove (6.2.2), by the transformation

$$y_i=x_i-\frac{\varphi_i(t)}{\varphi_n(t)}x_n\ (i=1,2,\cdots,n-1),\quad y_n=-\frac{1}{\varphi_n(t)}x_n$$

can be written as a linear system about $n-1$ unknown functions $y_1, y_2, \cdots, y_{n-1}$, not containing y_n.

8. Let $\boldsymbol{\Phi}(t)$ be a standard fundamental solution matrix ($\boldsymbol{\Phi}(0)=\boldsymbol{E}$) of $\boldsymbol{x}'=\boldsymbol{A}(t)\boldsymbol{x}$ ($\boldsymbol{A}$ is a $n\times n$ constant matrix), then prove

$$\boldsymbol{\Phi}(t)\boldsymbol{\Phi}^{-1}(t_0)=\boldsymbol{\Phi}(t-t_0)$$

where t_0 is a fixed number.

9. Suppose that $\boldsymbol{A}(t)$, $\boldsymbol{f}(t)$ are a continuous $n\times n$ matrix and an n-dimensional column vector on $[a,b]$, respectively. Prove that

$$\boldsymbol{x}'=\boldsymbol{A}(t)\boldsymbol{x}+\boldsymbol{f}(t)$$

has at most $n+1$ linear independent solutions.

10. Prove the superposition principle of linear inhomogeneous differential equations: If $\boldsymbol{x}_1(t)$, $\boldsymbol{x}_2(t)$ are solutions of

$$\boldsymbol{x}'=\boldsymbol{A}(t)\boldsymbol{x}+\boldsymbol{f}_1(t)$$
$$\boldsymbol{x}'=\boldsymbol{A}(t)\boldsymbol{x}+\boldsymbol{f}_2(t)$$

respectively, then $\boldsymbol{x}_1(t)+\boldsymbol{x}_2(t)$ is a solution of

$$\boldsymbol{x}'=\boldsymbol{A}(t)\boldsymbol{x}+\boldsymbol{f}_1(t)+\boldsymbol{f}_2(t)$$

11. Take $\boldsymbol{x}'=\boldsymbol{A}(t)\boldsymbol{x}+\boldsymbol{f}(t)$ into consideration, where

$$\boldsymbol{A}=\begin{bmatrix}2 & 1\\ 0 & 2\end{bmatrix},\ \boldsymbol{f}(t)=\begin{bmatrix}\sin t\\ \cos t\end{bmatrix}$$

(1) Check

$$\boldsymbol{\Phi}(t)=\begin{bmatrix}\mathrm{e}^{2t} & t\mathrm{e}^{2t}\\ 0 & \mathrm{e}^{2t}\end{bmatrix}$$

is a fundamental solution matrix of $\boldsymbol{x}'=\boldsymbol{A}(t)\boldsymbol{x}$.

(2) Find the solution $\boldsymbol{\varphi}(t)$ of $\boldsymbol{x}'=\boldsymbol{A}(t)\boldsymbol{x}+\boldsymbol{f}(t)$ with initial value $\boldsymbol{\varphi}(0)=\begin{bmatrix}1\\ -1\end{bmatrix}$.

12. Find the solution $\boldsymbol{\varphi}(t)$ of $\boldsymbol{x}'=\boldsymbol{A}(t)\boldsymbol{x}+\boldsymbol{f}(t)$ with initial value $\boldsymbol{\varphi}(0)=\begin{bmatrix}1\\ -1\end{bmatrix}$, where

$$\boldsymbol{A}=\begin{bmatrix}2 & 1\\ 0 & 2\end{bmatrix},\ \boldsymbol{f}(t)=\begin{bmatrix}0\\ \mathrm{e}^{2t}\end{bmatrix}.$$

13. Find the general solution of the system of equations

$$\begin{cases}\dfrac{\mathrm{d}x_1}{\mathrm{d}t}=x_1\cos^2 t-x_2(1-\sin t\cos t)\\ \dfrac{\mathrm{d}x_2}{\mathrm{d}t}=x_1(1+\sin t\cos t)+x_2\sin^2 t\end{cases}$$

with solutions $x_1=-\sin t$, $x_2=\cos t$.

14. Find the general solution of the system of equations

$$\begin{cases}\dfrac{\mathrm{d}x_1}{\mathrm{d}t}=\dfrac{1}{t}x_1-x_2+t\\ \\ \dfrac{\mathrm{d}x_2}{\mathrm{d}t}=\dfrac{1}{t^2}x_1-\dfrac{2}{t}x_2-t^2\end{cases},t>0$$

whose corresponding homogeneous system has solutions $x_1=t^2$, $x_1=-t$.

15. Given the system

$$\boldsymbol{x}'=\boldsymbol{A}(t)\boldsymbol{x} \tag{6.2.2}$$

where $\boldsymbol{A}(t)$ is a continuous $n\times n$ matrix on $[a,b]$, $\boldsymbol{\Phi}(t)$ is its fundamental solution matrix and n-dimensional vector function $\boldsymbol{F}(t,\boldsymbol{x})$ is continuous on $a\leqslant t\leqslant b$, $\|\boldsymbol{x}\|<\infty$. Try to show that the unique solution $\boldsymbol{\varphi}(t)$ of initial value problem

$$\begin{cases}\boldsymbol{x}'=\boldsymbol{A}(t)\boldsymbol{x}+\boldsymbol{F}(t,\boldsymbol{x})\\ \boldsymbol{\varphi}(t_0)=\boldsymbol{\eta},\ t_0\in[a,b]\end{cases} \tag{ * }$$

is a continuous solution of integral system

$$\boldsymbol{x}(t)=\boldsymbol{\Phi}(t)\boldsymbol{\Phi}^{-1}(t_0)\boldsymbol{\eta}+\int_{t_0}^{t}\boldsymbol{\Phi}(t)\boldsymbol{\Phi}^{-1}(s)\boldsymbol{F}(s,\boldsymbol{x}(s))\,\mathrm{d}s \tag{**}$$

On the contrary, the continuous solution of (**) is also the solution to initial value problem of (*).

16. In views of an $n\times n$ matrix $\boldsymbol{A}$, verify:

(1) For any constants c_1, c_2, it can be obtained

$$\exp(c_1\boldsymbol{A}+c_2\boldsymbol{A})=\exp c_1\boldsymbol{A}\cdot\exp c_2\boldsymbol{A}$$

(2) $$(\exp\boldsymbol{A})^k=\exp k\boldsymbol{A}$$

is correct for an arbitrary integer k. $(\exp\boldsymbol{A})^k=[(\exp\boldsymbol{A})^{-1}]^{-k}$ if k is a negative integer.

17. Show that

$$\boldsymbol{\varphi}(t)=\exp\boldsymbol{A}(t-t_0)\boldsymbol{\eta}$$

if $\boldsymbol{\varphi}(t)$ is the solution of $\boldsymbol{x}'=\boldsymbol{A}\boldsymbol{x}$ meeting the initial condition $\boldsymbol{\varphi}(t_0)=\boldsymbol{\eta}$.

18. Try to calculate the eigenvalues and their corresponding eigenvectors:

(1) $\begin{bmatrix}1 & 2\\ 4 & 3\end{bmatrix}$; (2) $\begin{bmatrix}2 & -3 & 3\\ 4 & -5 & 3\\ 4 & -4 & 2\end{bmatrix}$;

(3) $\begin{bmatrix}1 & 2 & 1\\ 1 & -1 & 1\\ 2 & 0 & 1\end{bmatrix}$; (4) $\begin{bmatrix}0 & 1 & 0\\ 0 & 0 & 1\\ -6 & -11 & -6\end{bmatrix}$.

19. Find a fundamental solution matrix of $\boldsymbol{x}'=\boldsymbol{A}\boldsymbol{x}$ and calculate $\exp \boldsymbol{A}t$, where $\boldsymbol{A}$ is:

(1) $\begin{bmatrix}-2 & 1\\ -1 & 2\end{bmatrix}$; (2) $\begin{bmatrix}1 & 2\\ 4 & 3\end{bmatrix}$;

(3) $\begin{bmatrix}2 & -3 & 3\\ 4 & -5 & 3\\ 4 & -4 & 2\end{bmatrix}$; (4) $\begin{bmatrix}1 & 0 & 3\\ 8 & 1 & -1\\ 5 & 1 & -1\end{bmatrix}$.

20. Give the fundamental solution matrix of $\boldsymbol{x}'=\boldsymbol{A}\boldsymbol{x}$ and the solution $\boldsymbol{\varphi}(t)$ with the initial value $\boldsymbol{\varphi}(0)=\boldsymbol{\eta}$:

(1) $\boldsymbol{A}=\begin{bmatrix}1 & 2\\ 4 & 3\end{bmatrix}$, $\boldsymbol{\eta}=\begin{bmatrix}3\\ 3\end{bmatrix}$;

(2) $\boldsymbol{A}=\begin{bmatrix}1 & 0 & 3\\ 8 & 1 & -1\\ 5 & 1 & -1\end{bmatrix}$, $\boldsymbol{\eta}=\begin{bmatrix}0\\ -2\\ -7\end{bmatrix}$;

(3) $\boldsymbol{A}=\begin{bmatrix}1 & 2 & 1\\ 1 & -1 & 1\\ 2 & 0 & 1\end{bmatrix}$, $\boldsymbol{\eta}=\begin{bmatrix}1\\ 0\\ 0\end{bmatrix}$.

21. Try to find out the solution $\boldsymbol{\varphi}(t)$ of $\boldsymbol{x}'=\boldsymbol{A}\boldsymbol{x}+\boldsymbol{f}(t)$:

(1) $\boldsymbol{\varphi}(0)=\begin{bmatrix}-1\\ 1\end{bmatrix}$, $\boldsymbol{A}=\begin{bmatrix}1 & 2\\ 4 & 3\end{bmatrix}$, $\boldsymbol{f}(t)=\begin{bmatrix}e^t\\ 1\end{bmatrix}$;

(2) $\boldsymbol{\varphi}(0)=\boldsymbol{0}$, $\boldsymbol{A}=\begin{bmatrix}0 & 1 & 0\\ 0 & 0 & 1\\ -6 & -11 & -6\end{bmatrix}$, $\boldsymbol{f}(t)=\begin{bmatrix}0\\ 0\\ e^{-t}\end{bmatrix}$;

(3) $\boldsymbol{\varphi}(0)=\begin{bmatrix}\eta_1\\ \eta_2\end{bmatrix}$, $\boldsymbol{A}=\begin{bmatrix}4 & -3\\ 2 & 1\end{bmatrix}$, $\boldsymbol{f}(t)=\begin{bmatrix}\sin t\\ -2\cos t\end{bmatrix}$.

22. Suppose that m is not a eigenvalue of the matrix $\boldsymbol{A}$. Show the linear inhomogeneous system of equations

$$\boldsymbol{x}'=\boldsymbol{A}\boldsymbol{x}+\boldsymbol{c}e^{mt}$$

has a solution of the form

$$\boldsymbol{x}(t)=\boldsymbol{p}\mathrm{e}^{mt}$$

where $\boldsymbol{c}$, $\boldsymbol{p}$ are constant vectors.

23. Consider the system

$$\begin{cases} x''_1-3x'_1+2x_1+x'_2-x_2=0 \\ x''_1-2x'_1+x'_2+x_2=0 \end{cases}$$

(1) Try to verify that the above system is equivalent to $\boldsymbol{u}'=\boldsymbol{A}\boldsymbol{u}$, where

$$\boldsymbol{u}=\begin{bmatrix} u_1 \\ u_2 \\ u_3 \end{bmatrix}=\begin{bmatrix} x_1 \\ x'_1 \\ x_2 \end{bmatrix}, \boldsymbol{A}=\begin{bmatrix} 0 & 1 & 0 \\ -4 & 4 & 2 \\ 2 & -1 & -1 \end{bmatrix}$$

(2) Give the fundamental solution matrix of the system in (1);

(3) Find the solution of the above system with the initial condition

$$x_1(0)=0,\ x'_1(0)=1,\ x_2(0)=0.$$

24. Find the solution to the following initial value problem:

(1) $\begin{cases} x'_1+x'_2=0, \\ x'_1-x'_2=1, \end{cases}$ $x_1(0)=1, x_2(0)=0;$

(2) $\begin{cases} x''_1+3x'_1+2x_1+x'_2+x_2=0, \\ x'_1+2x_1+x'_2-x_2=0, \end{cases}$ $x_1(0)=1, x'_1(0)=-1, x_2(0)=0;$

(3) $\begin{cases} x''_1-m^2x_2=0, \\ x''_2+m^2x_1=0, \end{cases}$ $x_1(0)=\eta_1, x'_1(0)=\eta_2,\ x_2(0)=\eta_3,\ x'_2(0)=\eta_4.$

25. If $y=\varphi(x)$ is the solution to the initial value problem of the second-order linear differential equation with the constant coefficients

$$\begin{cases} y''+ay'+by=0, \\ y(0)=0,\ y'(0)=1 \end{cases}$$

and then show that

$$y=\int_0^x \varphi(x-t)f(t)\,\mathrm{d}t$$

is the solution of the equation

$$y''+ay'+by=f(x)$$

where $f(x)$ is a continuous function.

Chapter 7 Qualitative[1] and stability[2] theories

The solution of differential equation can be obtained by elementary integral method, and its number is small. Therefore, it is necessary to study the properties of the solution according to the equation itself. Qualitative theory and stability theory were founded by French mathematician H. Poincare and Russian mathematician A. M. Lyapunov in the 1880s. These theories have very important theoretical and practical significance to the differential equation itself and other practical problems in the field of science and technology. So far, these theories continue to develop and have a lot of vitality. This chapter only briefly introduces some basic concepts and theories in this field, including the concept of phase plane[3], the classification of singularities of two-dimensional autonomous systems[4], examples of limit cycles, the concept of Lyapunov stability, the first approximation theory, and Lyapunov stability theorem.

7.1 Two-dimensional autonomous system and phase plane

The notation of Chapter 6 is still used

$$\boldsymbol{x}(t)=(x_1(t),x_2(t),\cdots,x_n(t))^{\mathrm{T}},(t,\boldsymbol{x})=(t,x_1,\cdots,x_n)$$

$$\boldsymbol{f}(t,\boldsymbol{x})=(f_1(t,\boldsymbol{x}),\cdots,f_n(t,\boldsymbol{x}))^{\mathrm{T}}$$

---

[1] qualitative 定性

[2] stability 稳定性

[3] phase plane 相平面

[4] two-dimensional autonomous systems 二维自治系统

Considering equations

$$\frac{d\boldsymbol{x}}{dt}=\boldsymbol{f}(t,\boldsymbol{x}) \tag{7.1.1}$$

If $\boldsymbol{f}$ does contain t, then the equations (7.1.1) are often referred to as nonautonomous systems[1]. If $\boldsymbol{f}$ does not contain t, that is,

$$\frac{d\boldsymbol{x}}{dt}=\boldsymbol{f}(\boldsymbol{x}) \tag{7.1.2}$$

Then (7.1.2) is an autonomous system[2]. In this chapter, we only study autonomous systems, in particular, in Section 7.1~7.3, we only study autonomous systems when $n=2$, when the system is often written as

$$\frac{dx}{dt}=X(x,y),\frac{dy}{dt}=Y(x,y) \tag{7.1.3}$$

where $X(x,y)$ and $Y(x,y)$ are functions of scalars x and y, which are continuous in a region G of the plane (x,y). Because X and Y do not contain t, they can be considered continuous in the region $I\times G$ of the space (t,x,y), where I denotes the interval $-\infty<t<+\infty$. Let the solution of the initial value problem of (7.1.3) be unique for any given initial value $(t_0,x_0,y_0)\in I\times G$. For example, if both $X(x,y)$ and $Y(x,y)$ satisfy the local Lipschitz condition for x and y, uniqueness can be guaranteed. The system (7.1.3) is commonly referred to as a two-dimensional autonomous system, or a planar autonomous system[3].

For example, differential equations describing the motion of a simple pendulum

$$\frac{d^2\theta}{dt^2}+\frac{k}{m}\frac{d\theta}{dt}+\frac{g}{l}\sin\theta=0$$

If $\frac{d\theta}{dt}=\omega$, then a system of equations is obtained.

$$\frac{d\theta}{dt}=\omega,\ \frac{d\omega}{dt}=-\frac{k}{m}\omega-\frac{g}{l}\sin\theta \tag{7.1.4}$$

This is a two-dimensional autonomous system.

Again, let the particle $M(x,y)$ move on the plane (x,y). In any instantaneous t, let the horizontal velocity v_x and vertical velocity v_y of the motion be the known functions $X(x,y)$ and $Y(x,y)$ of the position (x,y) of the particle M, respectively, both of which are independent of t, then the differential equation of motion of the particle M is

---

[1] nonautonomous systems 非自治系统

[2] autonomous system 自治系统

[3] planar autonomous system 平面自治系统

$$\frac{dx}{dt}=X(x,y),\ \frac{dy}{dt}=Y(x,y)$$

It is also a two-dimensional autonomous system.

Some basic concepts of planar autonomous systems are introduced below.

Similar to the first order equation, the solution of the system of equations (7.1.3)

$$x=x(t),\ y=y(t),\ t\in(\alpha,\beta) \tag{7.1.5}$$

The curve corresponding to space (t,x,y) becomes the integral curve❶ of (7.1.3). However, for autonomous systems (7.1.3), the following geometric explanations are usually used. Taking t as a parameter, the curve corresponding to the solution (7.1.5) on the plane (x,y) becomes the phase trajectory❷ of (7.1.3), or the trajectory for short, and defines the positive direction of the trajectory according to the increasing direction of t. A plane (x,y) is called a phase plane. The diagram consisting of all the orbits of (7.1.3) is called a phase diagram❸ of (7.1.3). Obviously, the trajectory of (7.1.3) on the phase plane (x,y) is the projection of the integral curve of (7.1.3) in space (t,x,y) on the plane (x,y), adding an arrow to the track in the direction of t to indicate the forward direction of the track.

If the two-dimensional autonomous system (7.1.3) is regarded as the velocity of the motion of the particle M in the region G defined on the plane (x,y), then it is obvious that the trajectory of the system is the trace of the motion of the particle, and the forward direction of the trajectory is the direction in which the particle moves. Therefore, from the point of view of dynamics, it is natural to use the trajectory on the phase plane to explain the solution of the two-dimensional autonomous system. It's more natural than an integral curve.

According to the initial assumption, the solution of the initial value problem of (7.1.3) exists and is unique, that is, for each point (t_0,x_0,y_0) in the region $I\times G$ in space (t,x,y), there is and only one integral curve passes through. Naturally, for each point (x_0,y_0) in the region G on phase space (x,y), is there only one trajectory of (7.1.3) passing through? The answer is yes. It is also because of this that it is possible to use trajectories and phase diagrams to illustrate the properties of the solution of the system (7.1.3).

Before demonstrating the above nature, give a concrete example.

【Example 7.1.1】 Give the solution of system

$$\frac{dx}{dt}=-y,\ \frac{dy}{dt}=x \tag{7.1.6}$$

---

❶ integral curve 积分曲线

❷ phase trajectory 相轨线

❸ phase diagram 相图

and draw the trajectory on the phase plane.

Solution. The solution satisfying the initial value condition $x(t_0)=x_0$, $y(t_0)=y_0$ is

$$\begin{cases} x=r_0\cos(t-t_0+\varphi_0) \\ y=r_0\sin(t-t_0+\varphi_0) \end{cases} \tag{7.1.7}$$

Where $r_0\cos\varphi_0=x_0$, $r_0\sin\varphi_0=y_0$. If x_0 and y_0 are not zero at the same time, then it is known from (7.1.7) that the integral curve passing through point (t_0, x_0, y_0) is a spiral curve located on the cylindrical surface

$$x^2+y^2=r_0^2$$

in space (t, x, y). On the phase plane (x, y), the corresponding trajectory of (7.1.7) can be obtained by eliminating t, which is a circumference

$$x^2+y^2=r_0^2 \tag{7.1.8}$$

on the plane (x, y), and the forward direction of the trajectory is counterclockwise. If $x_0=y_0=0$, i. e. $r_0=0$, then it is known by (7.1.7) that the integral curve passing through $(t_0, 0, 0)$ is a straight line $x=0$, $y=0$ in space (t, x, y), that is, the t axis. On the phase plane (x, y), the trajectory of (7.1.7) corresponds to a point $x=0$, $y=0$.

From the above discussion, it can be seen that for given x_0, y_0, there is only one orbital line (7.1.8) for (7.1.6) passing through the point (x_0, y_0) on the phase plane (x, y). The integral curve of (7.1.6) and the trajectory on the phase plane are drawn in Figs. 7.1.1 and 7.1.2, respectively.

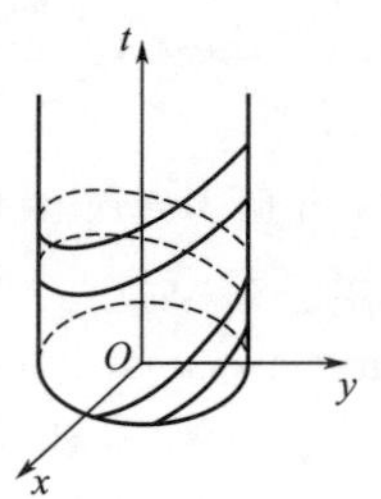

Figure 7.1.1.

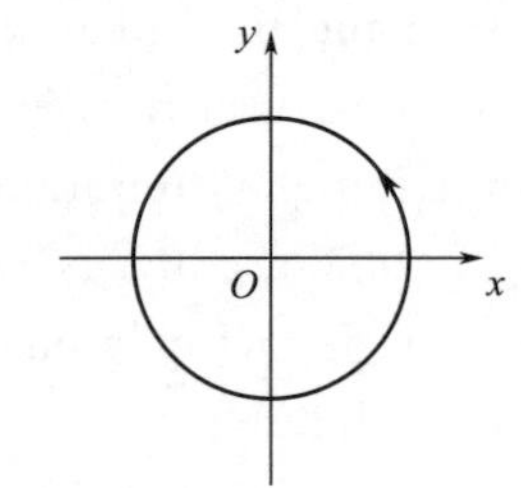

Figure 7.1.2.

Some important properties of solutions and trajectories of the autonomous system (7.1.3) are proved below. Here, the maximum existence interval of each solution for (7.1.3) is $(-\infty, +\infty)$.

Lemma 7.1.1. Let $x(t)$, $y(t)$ be a solution of the autonomous system (7.1.3), then for any given α, $x(t-\alpha)$, $y(t-\alpha)$ is also a solution of (7.1.3). And the corresponding orbits of the two solutions coincide with each other.

Proof. Because $x(t)$, $y(t)$ is the solution of (7.1.3), there is an identity for any τ

$$\begin{cases} \dfrac{dx(\tau)}{d\tau}=X(x(\tau),\ y(\tau)) \\ \dfrac{dy(\tau)}{d\tau}=X(x(\tau),\ y(\tau)) \end{cases}$$

Let $\tau=t-\alpha$ and replace it with the above formula, notice $d\tau=d(t-\alpha)=dt$, so there is

$$\begin{cases} \dfrac{dx(t-\alpha)}{dt}=X(x(t-\alpha),\ y(t-\alpha)) \\ \dfrac{dy(t-\alpha)}{dt}=X(x(t-\alpha),\ y(t-\alpha)) \end{cases}$$

This shows that function $x=x(t-\alpha)$, $y=y(t-\alpha)$ is also the solution of (7. 1. 3). Obviously, the trajectory of solution $x=x(t)$, $y=y(t)$ coincides with that of solution $x=x(t-\alpha)$, $y=y(t-\alpha)$, and the positive direction is the same, but the corresponding t values of the points on the trajectory are different for different solutions. For example, let the first solution $x=x(t)$, $y=y(t)$ correspond to a point (x_0, y_0) on the trajectory at $t=t_0$, in other words, $x(t_0)=x_0$, $y(t_0)=y_0$, then the second solution $x=x(t-\alpha)$, $y=y(t-\alpha)$ corresponds to this point at $t=t_0+\alpha$.

Lemma 7. 1. 2. Any two orbits of an autonomous system (7. 1. 3) either do not intersect or coincide completely, that is, through any point (x_0, y_0) in the region G on the phase plane (x, y), the autonomous system (7. 1. 3) has and has only one trajectory.

This property is called "uniqueness of the trajectory[1]".

Proof. Let $x=x_1(t)$, $y=y_1(t)$ be a solution of (7. 1. 3), which satisfies condition $x_1(t_1)=x_0$, $y_1(t_1)=y_0$ and its corresponding trajectory is L_1. Let $x=x_2(t)$, $y=y_2(t)$ also be a solution of (7. 1. 3), which satisfies condition $x_2(t_2)=x_0$, $y_2(t_2)=y_0$ and its corresponding trajectory is L_2. L_1 and L_2 have common point (x_0, y_0). It is proved that L_1 and L_2 coincide completely.

In fact, in the second solution, replace t with $t-t_1+t_2$, and get

$$x=x_2(t-t_1+t_2),\ y=y_2(t-t_1+t_2) \tag{7. 1. 9}$$

Known by Lemma 7. 1. 1, (7. 1. 9) is also the solution of (7. 1. 3), and it satisfies

$$\begin{cases} x\big|_{t=t_1}=x_2(t-t_1+t_2)=x_0 \\ y\big|_{t=t_1}=y_2(t-t_1+t_2)=y_0 \end{cases} \tag{7. 1. 10}$$

It is noted that $x=x_1(t)$, $y=y_1(t)$ is the solution of (7. 1. 3) and satisfies the same initial value condition (7. 1. 10). Therefore, the uniqueness theorem (Theorem 6. 1. 1) of the solution of the initial value problem gives that

---

[1] uniqueness of the trajectory 轨线的唯一性

$$\begin{cases} x_2(t-t_1+t_2)=x_1(t) \\ y_2(t-t_1+t_2)=y_1(t) \end{cases}$$

However, it is known from Lemma 7.1.1 that the trajectory corresponding to solution $x_2(t-t_1+t_2)$, $y_2(t-t_1+t_2)$ coincides with that corresponding to solution $x_2(t)$, $y_2(t)$, so the trajectories of solutions $x_1(t)$, $y_1(t)$ and $x_2(t)$, $y_2(t)$ coincide with each other.

Remark 7.1.1. It can be seen from the above proof that if the trajectories corresponding to the two solutions have a common point, one of the solutions can be obtained by the translation of the other one by t. Combined with Lemma 7.1.1, we have

Lemma 7.1.3. The necessary and sufficient condition for the trajectory coincidence of two solutions of an autonomous system (7.1.3) is that one of the solutions can be obtained by the proper translation of the other one by t.

It is not difficult for readers to deepen their understanding of the above Lemma through Example 7.1.1.

Now introduce the notation. Let $x(t,t_0,x_0,y_0)$, $y(t,t_0,x_0,y_0)$ denote the solution of (7.1.3) satisfying condition $x|_{t=t_0}=x_0, y|_{t=t_0}=y_0$. Apparently,

$$x(t,t_0,x_0,y_0)=x_0,\quad y(t,t_0,x_0,y_0)=y_0.$$

Set the trajectory L and point $(x_0,y_0)\in L$. If (t,t_0,x_0,y_0), $y(t,t_0,x_0,y_0)$ is a solution corresponding to L, then

$$x=x(t+t_0,t_0,x_0,y_0),\quad y=y(t+t_0,t_0,x_0,y_0) \tag{7.1.11}$$

is also a solution corresponding to L, and it satisfies condition $x|_{t=t_0}=x_0$, $y|_{t=t_0}=y_0$. So (7.1.11) can be written as

$$\begin{cases} x(t+t_0,t_0,x_0,y_0)=x(t,0,x_0,y_0) \\ y(t+t_0,t_0,x_0,y_0)=y(t,0,x_0,y_0) \end{cases}$$

So the solution corresponding to the trajectory L through point (x_0,y_0) can always be written as $x(t,0,x_0,y_0)$, $y(t,0,x_0,y_0)$, abbreviated as $x(t,x_0,y_0)$, $y(t,x_0,y_0)$.

Lemma 7.1.4. If

$$\begin{cases} x(t_1,x_0,y_0)=x_1, y(t_1,x_0,y_0)=y_1 & (7.1.12) \\ x(t_2,x_0,y_0)=x_2, y(t_2,x_0,y_0)=y_2 & (7.1.13) \end{cases}$$

Then

$$\begin{cases} x(t_1+t_2,x_0,y_0)=x(t_2,x(t_1,x_0,y_0),y(t_1,x_0,y_0)) \\ y(t_1+t_2,x_0,y_0)=y(t_2,x(t_1,x_0,y_0),y(t_1,x_0,y_0)) \end{cases} \tag{7.1.14}$$

Proof. It is known from Lemma 7.1.1 that

$$x=x(t-t_1,x_1,y_1),\quad y=y(t-t_1,x_1,y_1)$$

is also the solution of (7.1.3) and satisfies $x|_{t=t_1}=x_1$, $y|_{t=t_1}=y_1$. The solution

$$x=x(t,x_0,y_0),\ y=y(t,x_0,y_0)$$

of (7.1.3) satisfies the same condition from (7.1.12). The unique Theorem of the solution of the initial value problem tells us that substituting $t=t_1+t_2$ for

$$\begin{cases}x(t,x_0,y_0)=x(t-t_1,x_1,y_1),\\ y(t,x_0,y_0)=y(t-t_1,x_1,y_1).\end{cases} \tag{7.1.15}$$

gets (7.1.14).

Now let's talk about the meaning of Lemma 7.1.4. The trajectory of (7.1.3) through (x_0,y_0) at $t=0$ passes through (x_1,y_1) at $t=t_1$; the trajectory of (7.1.3) through (x_1,y_1) at $t=0$ passes through (x_2,y_2) at $t=t_2$, therefore, the trajectory of (7.1.3) through (x_0,y_0) at $t=0$ goes through (x_2,y_2) at $t=t_1+t_2$.

Remark 7.1.2. For a fixed t, the solution

$$x=x(t,x_0,y_0),\ y=y(t,x_0,y_0) \tag{7.1.16}$$

of (7.1.3) can be regarded as a transformation from (x_0,y_0) to (x,y) of the phase plane. For $t\in(-\infty,+\infty)$, (7.1.16) is a family of transformations depending on the parameter t. For simplicity, p_0, p denote (x_0,y_0), (x,y), and f_t is the transformation. So (7.1.16) can be simply written as $f_tp_0=p$. It is easy to know that the following conclusions hold:

(1) $f_0p_0=p_0$;

(2) $f_\lambda f_tp_0=f_{\lambda+t}p_0$;

(3) $f_{-t}f_tp_0=f_0p_0$.

Here (1) and (3) are obvious, and (2) is Lemma 7.1.4. Because f_t has the above properties, f_t constitutes a transformation group as $t\in(-\infty,+\infty)$. Abstract dynamic system is developed from this concept.

It can also be seen from Example 7.1.1 that the orbit is not necessarily a "line", but can be a point. For example, $x=0$, $y=0$ is a solution of (7.1.6), its corresponding trajectory on the phase plane is a point (0,0). This kind of trajectory is very important. We have the following definition.

Definition 7.1.1. A point that holds

$$X(x,y)=0,\ Y(x,y)=0$$

simultaneously on a phase plane is called a singularity[1] of (7.1.3).

There is obviously the following Lemma.

Lemma 7.1.5. Let (x_0,y_0) be a singularity of (7.1.3), then $x=x_0$, $y=y_0$ is a

---

[1] singularity 奇点

solution of (7.1.3) (this solution is called a constant solution[1]); conversely, if $x=x_0$, $y=y_0$ is a constant solution of (7.1.3), then the point (x_0, y_0) on the phase plane is a singularity of (7.1.3). Singularity (x_0, y_0) is the trajectory corresponding to constant solution $x=x_0$, $y=y_0$.

Proof. Let (x_0, y_0) be the singularity of (7.1.3), then

$$X(x_0, y_0)=Y(x_0, y_0)=0$$

Therefore, it is known that the left and right sides of (7.1.3) with $x=x_0$, $y=y_0$ become identical. So $x=x_0$, $y=y_0$ is a solution of (7.1.3). On the contrary, $x=x_0$, $y=y_0$ is a constant solution of (7.1.3), after substituting that into (7.1.3) it is an identity:

$$0=\frac{dx_0}{dt}=X(x_0, y_0),\quad 0=\frac{dy_0}{dt}=Y(x_0, y_0)$$

So it is known that (x_0, y_0) is the singularity of (7.1.3). In addition, it is clear that the trajectory corresponding to the constant solution $x=x_0$, $y=y_0$ is only one point, that is, singularity (x_0, y_0).

Another kind of trajectory, such as (7.1.8), can also be seen from Example 7.1.1, which is a closed curve. Its corresponding solution (7.1.7) is a periodic solution. This kind of trajectory which forms a closed curve is called a closed orbit[2]. Closed orbits and singularities are also very important orbits.

Lemma 7.1.6. The trajectory of an autonomous system (7.1.3) is a closed orbit if and only if its corresponding solution is a non-constant periodic solution (Certainly, the constant solution can also be considered as a periodic solution. And here we consider non-constant periodic solutions).

Proof. Adequacy. Let $x=x(t)$, $y=y(t)$ be a non-constant periodic solution of (7.1.3), then there exists $T>0$ such that for any t, we have

$$x(t+T)=x(t),\quad y(t+T)=y(t)$$

This means that the point $(x(t_0), y(t_0))$ corresponding to any t_0 on the trajectory returns to this point at t_0+T. But not for all t, $x(t)=x(t_0)$, $y(t)=y(t_0)$. This shows that the trajectory is a closed curve, that is, a closed orbit.

Necessity. Assume Γ is a closed orbit of (7.1.3) and take any point (x_0, y_0) on Γ. $x(t, x_0, y_0)$, $y(t, x_0, y_0)$ is a solution corresponding to Γ of (7.1.3) satisfying condition $x|_{t=0}=x_0$, $y|_{t=0}=y_0$. Because Γ is a closed orbit, let the point move in the increasing direction of t until arriving at (x_0, y_0) at the second time, at this time, $t=T$, where T is a positive constant. That means that

[1] constant solution 常数解

[2] closed orbit 闭轨

$$\begin{cases} x(T, x_0, y_0) = x(0, x_0, y_0) = x_0 \\ y(T, x_0, y_0) = y(0, x_0, y_0) = y_0 \end{cases}$$

Therefore, from the formula (7.1.14) of Lemma 7.1.4, it is deduced that for any t,

$$\begin{cases} x(t+T, x_0, y_0) = x(t, x(T, x_0, y_0), y(T, x_0, y_0)) = x(t, x_0, y_0) \\ y(t+T, x_0, y_0) = y(t, x(T, x_0, y_0), y(T, x_0, y_0)) = y(t, x_0, y_0) \end{cases}$$

It is proved that the solution $x(t, x_0, y_0)$, $y(t, x_0, y_0)$ is a periodic solution with the period T. This periodic solution is certainly not a constant solution, because the constant solution corresponds only to singularities.

In addition, it is known from Lemma 7.1.3 that any solution corresponding to Γ must be written as $x(t-\alpha, x_0, y_0)$, $y(t-\alpha, x_0, y_0)$, where α is the corresponding constant, and obviously it is also a periodic solution with a period of T.

In addition to singularities and closed orbits, an autonomous system (7.1.3) has another trajectory (not seen in Example 7.1.1), that is, an ordinary open arc, known by Lemma 7.1.2, which does not intersect itself. This open arc is called an open orbit[1]. Therefore, the trajectory of autonomous system can be divided into three kinds: singularity, closed orbit and open orbit. The following two sections describe some of the basics of singularities and closed orbits, respectively.

7.2 Plane singularity[2]

This section examines the behavior of the trajectory of two-dimensional autonomous systems

$$\frac{dx}{dt} = X(x, y), \quad \frac{dy}{dt} = Y(x, y) \tag{7.2.1}$$

in its singularity domain.

Let the origin $O(0,0)$ be one of its singularities, that is,

$$X(0,0) = 0, \quad Y(0,0) = 0 \tag{7.2.2}$$

and $X(x, y)$ and $Y(x, y)$ have continuous 1st-order partial derivatives in a neighborhood of point O. Then according to Taylor's formula, $X(x, y)$ and $Y(x, y)$ at O can be expanded to

$$X(x, y) = ax + by + R_1(x, y)$$
$$Y(x, y) = cx + dy + R_2(x, y)$$

where

---

[1] open orbit 开轨线

[2] plane singularity 平面奇点

$$\begin{bmatrix} a & b \\ c & d \end{bmatrix} = \begin{bmatrix} \dfrac{\partial X}{\partial x} & \dfrac{\partial X}{\partial y} \\ \dfrac{\partial Y}{\partial x} & \dfrac{\partial Y}{\partial y} \end{bmatrix}_{x=0,\ y=0} \tag{7.2.3}$$

$$\lim_{r\to 0} \frac{R_i(x,y)}{r} = 0\ (i=1,\ 2),\ r=\sqrt{x^2+y^2} \tag{7.2.4}$$

Substitute the expansions of $X(x,y)$ and $Y(x,y)$ into the right of (7.2.1) to get

$$\begin{cases} \dfrac{\mathrm{d}x}{\mathrm{d}t} = ax+by+R_1(x,y) \\ \dfrac{\mathrm{d}y}{\mathrm{d}t} = cx+dy+R_2(x,y) \end{cases} \tag{7.2.1$'$}$$

In other words, under the given conditions, system (7.2.1) can be written as (7.2.1)$'$ in the domain of singularity O.

The system

$$\begin{cases} \dfrac{\mathrm{d}x}{\mathrm{d}t} = ax+by \\ \dfrac{\mathrm{d}y}{\mathrm{d}t} = cx+dy \end{cases} \tag{7.2.5}$$

corresponding to the linear part of (7.2.1)$'$ is called the first approximation system[1] of (7.2.1)$'$.

Next, we investigate (7.2.5), followed by (7.2.1)$'$. Finally, it is extended to the general singularity (x_0, y_0).

7.2.1 Trajectory distribution of two-dimensional linear systems

The system (7.2.5) is a system of second-order homogeneous linear equations with constant coefficients, and its solution is introduced in detail in Chapter 6. Now, according to the method of the chapter, the general solution of (7.2.5) is obtained, then the characteristics of its phase diagram are drawn, and the singularities are classified.

The coefficient matrix of the system (7.2.5) is represented by $\boldsymbol{A}$, that is,

$$\boldsymbol{A} = \begin{bmatrix} a & b \\ c & d \end{bmatrix} \tag{7.2.6}$$

So that (7.2.5) can be written as

$$\frac{\mathrm{d}}{\mathrm{d}t}\begin{bmatrix} x \\ y \end{bmatrix} = \boldsymbol{A}\begin{bmatrix} x \\ y \end{bmatrix} \tag{7.2.5$'$}$$

The characteristic equation of $\boldsymbol{A}$ is

---

[1] first approximation system 一次近似系统

$$\det(\boldsymbol{A}-\lambda \boldsymbol{I})=\begin{vmatrix} a-\lambda & b \\ c & d-\lambda \end{vmatrix}=0$$

Denote $p=-(a+d)$, $q=ad-bc$, then the characteristic equation can be written as.

$$\lambda^2+p\lambda+q=0 \tag{7.2.7}$$

Let's keep setting $q\neq 0$, that is, (7.2.7) has no zero root. The corresponding singularity $O(0,0)$ is called an elementary singularity[1]. According to the different cases of characteristic roots λ_1, λ_2, we can divide into three categories.

1. λ_1 and λ_2 are a pair of unequal real roots.

At this time, Jordan standard form of matrix $\boldsymbol{A}$ is.

$$\boldsymbol{J}_1=\begin{bmatrix} \lambda_1 & 0 \\ 0 & \lambda_2 \end{bmatrix}$$

Let $\boldsymbol{v}_1$, $\boldsymbol{v}_2$ be the eigenvectors of $\boldsymbol{A}$ belonging to λ_1, λ_2, and it is known by linear algebra theory

$$\boldsymbol{T}_1^{-1}\boldsymbol{A}\boldsymbol{T}_1=\boldsymbol{J}_1$$

where $\boldsymbol{T}_1=(\boldsymbol{v}_1, \boldsymbol{v}_2)$.

Linear exchange

$$\begin{bmatrix} x \\ y \end{bmatrix}=\boldsymbol{T}_1\begin{bmatrix} \xi \\ \eta \end{bmatrix}=(\boldsymbol{v}_1, \boldsymbol{v}_2)\begin{bmatrix} \xi \\ \eta \end{bmatrix} \tag{7.2.8}$$

makes (7.2.5) be $\boldsymbol{T}_1\dfrac{\mathrm{d}}{\mathrm{d}t}\begin{bmatrix} \xi \\ \eta \end{bmatrix}=\boldsymbol{A}\boldsymbol{T}_1\begin{bmatrix} \xi \\ \eta \end{bmatrix}$, that is

$$\frac{\mathrm{d}}{\mathrm{d}t}\begin{bmatrix} \xi \\ \eta \end{bmatrix}=\boldsymbol{T}_1^{-1}\boldsymbol{A}\boldsymbol{T}_1\begin{bmatrix} \xi \\ \eta \end{bmatrix}=\boldsymbol{J}_1\begin{bmatrix} \xi \\ \eta \end{bmatrix} \tag{7.2.9}$$

The general solution of (7.2.9) is

$$\begin{bmatrix} \xi \\ \eta \end{bmatrix}=\begin{bmatrix} C_1\mathrm{e}^{\lambda_1 t} \\ C_2\mathrm{e}^{\lambda_2 t} \end{bmatrix} \tag{7.2.10}$$

It is now discussed on plane (ξ, η). We can divide into three subclasses.

(1) $\lambda_1\lambda_2<0$. Let $\lambda_1<0$, $\lambda_2>0$ (otherwise, exchange ξ and η).

Obviously, if and only if $C_1=C_2=0$, the trajectory (7.2.10) is a point $O(0,0)$, that is, the singularity. The trajectories discussed below no longer contain O.

If $C_1\neq 0=C_2$, the trajectory coincides with ξ axis, and $\xi\to 0$ as $t\to+\infty$.

If $C_1=0\neq C_2$, the trajectory coincides with η axis, and $\eta\to 0$ as $t\to-\infty$.

If $C_1C_2\neq 0$, eliminating t from (7.2.10) to get $\left(\dfrac{\xi}{C_1}\right)^{\lambda_2}=\left(\dfrac{\eta}{C_2}\right)^{\lambda_1}$, i.e.,

---

[1] elementary singularity 初等奇点

$$\left(\frac{\xi}{C_1}\right)^{\lambda_2}\left(\frac{\eta}{C_2}\right)^{-\lambda_1}=1$$

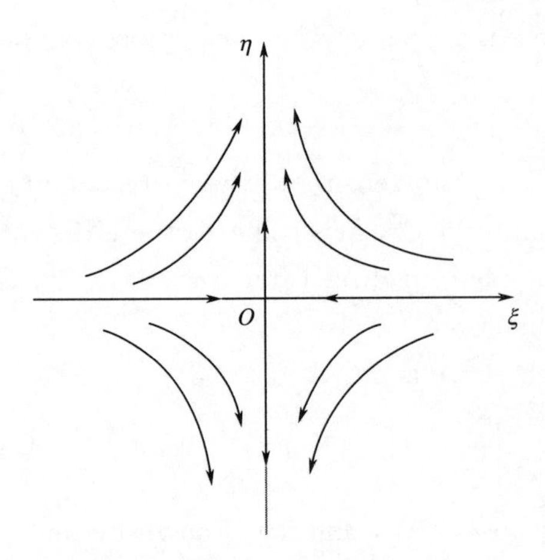

Figure 7.2.1.

It looks like a family of "hyperbolic curves". As can be seen from (7.2.10), $\xi\to 0$, $\eta\to\infty$ as $t\to+\infty$. As a result, the forward direction of the trajectory can be drawn. Figure 7.2.1 shows the orbits of (7.2.10) on plane (ξ,η). This singularity O becomes the saddle point❶. The characteristic of saddle point is that it has a neighborhood in which just two orbits tend to this singularity along a pair of opposite directions as $t\to+\infty$; in addition, just two more orbits tend to this singularity in a pair of opposite directions as $t\to-\infty$. With the exception of this singularity and the above trajectories, all the orbits in the neighborhood will leave the neighborhood when t increases and decreases.

(2) $\lambda_1<0$, $\lambda_2<0$. Let $\lambda_2<\lambda_1<0$ (otherwise, exchange ξ and η).

The situation can be modelled on (1) when $C_1C_2=0$.

If $C_1C_2\neq 0$, eliminating t from (7.2.10) to get

$$\frac{\eta}{C_2}=\left(\frac{\xi}{C_1}\right)^{\frac{\lambda_2}{\lambda_1}}$$

where $\frac{\lambda_2}{\lambda_1}>1$. The orbits are like a family of "parabola", however, as is known from (7.2.10), point O should be dug out of it. The corresponding points on all orbits tend to $O(0,0)$ and tangent to ξ axis at point O as $t\to+\infty$. Figure 7.2.2 shows orbits of (7.2.10) on the plane (ξ,η).

This singularity O is called a stable node❷.

(3) $\lambda_1>0$, $\lambda_2>0$. Let $\lambda_2>\lambda_1>0$ (otherwise, exchange ξ and η).

This situation is equivalent to replacing t with $-t$ in (2). Singularity O is called an unstable node❸, see Figure 7.2.3.

The characteristic of a stable (unstable) node is that it has a neighborhood in which all orbits except the singularity tend to the singularity as $t\to+\infty$ $(t\to-\infty)$,

❶ saddle point 鞍点

❷ stable node 稳定结点

❸ unstable node 不稳定结点

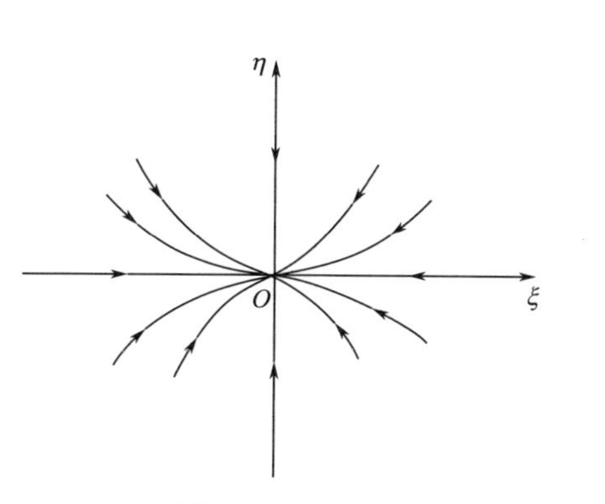

Figure 7. 2. 2.

Figure 7. 2. 3.

and two of them tend to the singularity in a pair of opposite directions, however, the others tend to this singularity in the other pair.

【Example 7. 2. 1】 Determine the singularity type of the system

$$\begin{cases} \dfrac{dx}{dt}=4x+y \\ \dfrac{dy}{dt}=2x+5y \end{cases}$$

Solution. The characteristic equation of coefficient matrix $\boldsymbol{A}=\begin{bmatrix} 4 & 1 \\ 1 & 5 \end{bmatrix}$ is

$$\begin{vmatrix} 4-\lambda & 1 \\ 2 & 5-\lambda \end{vmatrix}=\lambda^2-8\lambda+18=0$$

The characteristic roots $\lambda_1=3$, $\lambda_2=6$ are a pair of unequal real roots, and all of them are greater than zero, so singularity O is an unstable node.

2. λ_1 and λ_2 are a pair of conjugate complex roots $\lambda_1=\alpha+i\beta$, $\lambda_2=\alpha-i\beta$, $\beta>0$.

At this point, the Jordan canonical form of matrix $\boldsymbol{A}$ is

$$\boldsymbol{J}_2=\begin{bmatrix} \alpha & \beta \\ -\beta & \alpha \end{bmatrix}, \ \beta>0$$

It is known from linear algebra that $\boldsymbol{A}$ can be transformed into its Jordan canonical form by a nonsingular linear transformation $\boldsymbol{T}_2$:

$$\boldsymbol{T}_2^{-1}\boldsymbol{A}\boldsymbol{T}_2=\boldsymbol{J}_2$$

At the same time, assume that $\begin{bmatrix} x \\ y \end{bmatrix}=\boldsymbol{T}_2\begin{bmatrix} \xi \\ \eta \end{bmatrix}$, (7. 2. 5) becomes

$$\frac{d}{dt}\begin{bmatrix} \xi \\ \eta \end{bmatrix}=\boldsymbol{J}_2\begin{bmatrix} \xi \\ \eta \end{bmatrix}$$

Its general solution

$$\begin{cases}\xi = Ce^{\alpha t}\cos(\varphi - \beta t)\\ \eta = Ce^{\alpha t}\sin(\varphi - \beta t)\end{cases} \quad (7.2.11)$$

can be obtained, where C and φ are arbitrary constants.

Introducing polar coordinates (r, θ) into the plane (ξ, η):

$$\xi = r\cos\theta,\ \eta = \sin\theta$$

so (7.2.11) gives

$$r = Ce^{\alpha t},\ \theta = -\beta t + \varphi \quad (7.2.12)$$

The trajectory corresponding to constant $C=0$ is a singularity $O(0,0)$, and the corresponding trajectory for $C \neq 0$ can be divided into three subclasses:

(1) $\alpha = 0$, i.e., $\lambda_{1,2} = \alpha \pm i\beta$.

(7.2.12) becomes $r = C$, $\theta = -\beta t + \varphi$. So the trajectory on the plane (ξ, η) is a family of concentric circles. $\theta \to -\infty$ as $t \to +\infty$, that is, the orbits rotate clockwise on the plane (ξ, η). The trajectory on the plane (ξ, η) is shown in Figure 7.2.4.

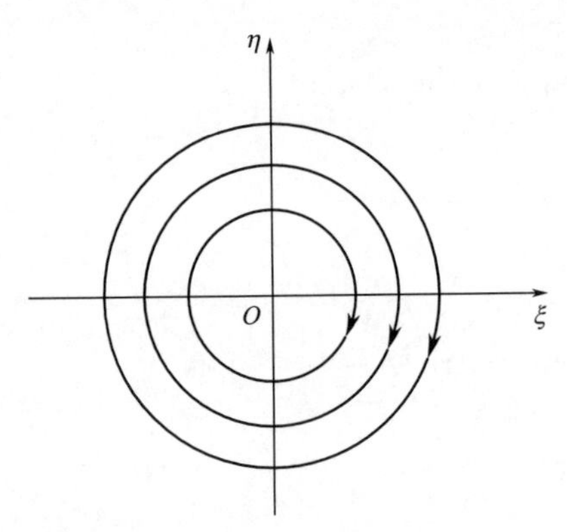

Figure 7.2.4.

This singularity becomes the center❶. The characteristic of the center is that it has a neighborhood, and all orbits except the singularity in the neighborhood are closed curves around the singularity.

(2) $\alpha < 0$.

Eliminating t from the two formulas in (7.2.12), the trajectory in the plane (ξ, η) is a family of spirals

$$r = Ce^{-\frac{\alpha}{\beta}(\theta - \varphi)}$$

around the origin. $\theta \to -\infty$ and $r \to 0$ as $t \to +\infty$. That is, it is clockwise (that is, negative direction) rotation to point O, shown in Figure 7.2.5.

This singularity O is called a stable focus. ❷

(3) $\alpha > 0$.

This situation is equivalent to replacing t with $-t$ in (2), thus the trajectory in the plane (ξ, η) is still a family of spirals, and it is counterclockwise rotation to point O as $t \to -\infty$, as shown in Figure 7.2.6. Singularity O is called an unstable focus❸.

---

❶ center 中心

❷ stable focus 稳定焦点

❸ unstable focus 不稳定焦点

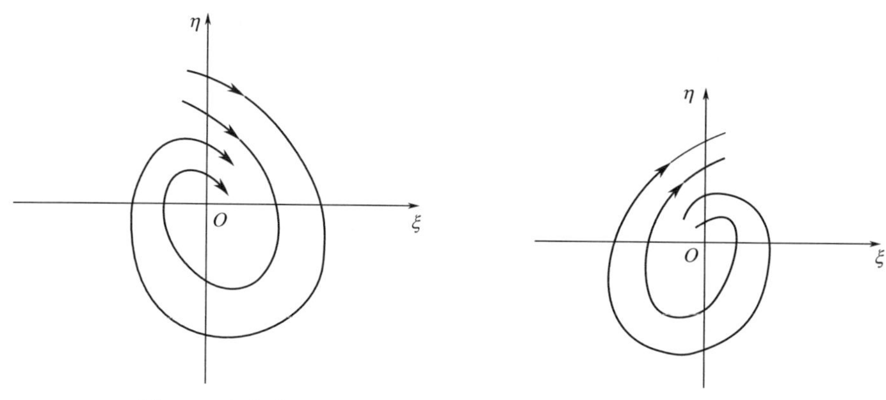

Figure 7.2.5.　　　　Figure 7.2.6.

The characteristic of the stable (unstable) focus is that it has a neighborhood through which all orbits except the singularity rotate infinitely around the singularity and tend to the singularity at $t \to +\infty$ ($t \to -\infty$).

【Example 7.2.2】 Determine the singularity type of the system

$$\begin{cases} \dfrac{dx}{dt} = -3x - 2y \\ \dfrac{dy}{dt} = 5x - y \end{cases}$$

Solution. The characteristic equation of coefficient matrix $\boldsymbol{A} = \begin{bmatrix} -3 & -2 \\ 5 & -1 \end{bmatrix}$ is

$$\begin{vmatrix} -3-\lambda & -2 \\ 5 & -1-\lambda \end{vmatrix} = \lambda^2 + 4\lambda + 13 = 0$$

The characteristic roots $\lambda_{1,2} = -2 \pm 3i$ are a pair of conjugate complex roots, whose real parts $-2 < 0$, so singularity O is a stable focus.

3. $\lambda_1 = \lambda_2$ is a double real root.

At this point, there are two kinds of Jordan standard forms of $\boldsymbol{A}$, which are recorded as $\boldsymbol{J}_3$ and $\boldsymbol{J}_4$.

(1) $\boldsymbol{J}_3 = \begin{bmatrix} \lambda_1 & 0 \\ 0 & \lambda_1 \end{bmatrix}$.

If $b = c = 0$, $a = d$, Jordan standard form of $\boldsymbol{A}$ is $\boldsymbol{J}_3$ and $\lambda_1 = a$. On the contrary, when Jordan standard form of $\boldsymbol{A}$ is $\boldsymbol{J}_3$, there exists a second-order nonsingular matrix $\boldsymbol{T}_3$ such that

$$\boldsymbol{T}_3^{-1} \boldsymbol{A} \boldsymbol{T}_3 = \boldsymbol{J}_3$$

therefore,

$$\boldsymbol{A} = \boldsymbol{T}_3 \boldsymbol{J}_3 \boldsymbol{T}_3^{-1} = \lambda_1 \boldsymbol{T}_3 \boldsymbol{I} \boldsymbol{T}_3^{-1} = \lambda_1 \boldsymbol{T}_3 \boldsymbol{T}_3^{-1} = \lambda_1 \boldsymbol{I} = \boldsymbol{J}_3$$

that is,

$$b=c=0,\ a=d=\lambda_1$$

In short, for this case, the system (7.2.5) is

$$\frac{\mathrm{d}}{\mathrm{d}t}\begin{bmatrix}x\\y\end{bmatrix}=\begin{bmatrix}\lambda_1 & 0\\0 & \lambda_1\end{bmatrix}\begin{bmatrix}x\\y\end{bmatrix}$$

whose general solution is

$$x=C_1\mathrm{e}^{\lambda_1 t},\ y=C_2\mathrm{e}^{\lambda_1 t} \tag{7.2.13}$$

where C_1 and C_2 are arbitrary constants.

If $C_1=C_2=0$, the trajectory is the singularity $O(0,0)$.

If $C_1^2+C_2^2\neq 0$, (7.2.13) is a family of rays with point O. as the endpoint, but does not contain point O. $\lambda_1<0$ gives that all orbits tend to singularity O as $t\to+\infty$, which is called a stable critical node[1]. All orbits tend to O as $t\to-\infty$ when $\lambda_1>0$, this singularity O is called an unstable critical node[2]. Figures 7.2.7 and 7.2.8 draw the trajectories of these cases on the plane (x,y), respectively.

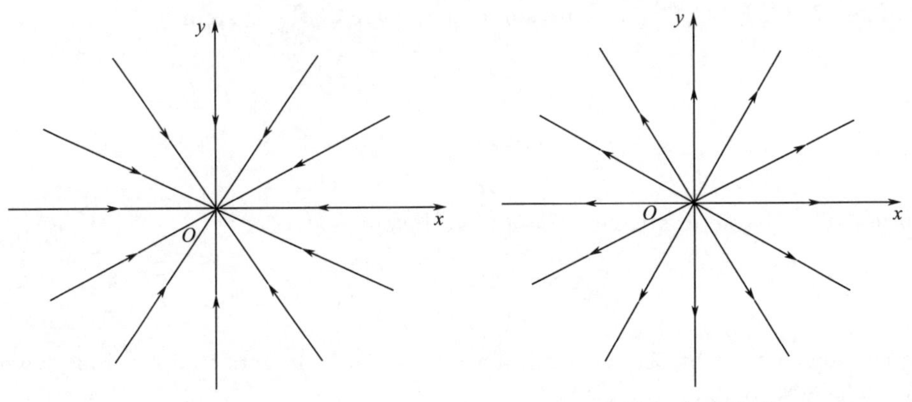

Figure 7.2.7. Figure 7.2.8.

The characteristics of the stable (unstable) critical node are as follows: it has a neighborhood in which all the orbits except the singularity tend to the singularity as $t\to+\infty$ ($t\to-\infty$), and when each trajectory tends to the singularity, they follow a certain direction respectively; And only one trajectory in each direction tends to singularity along it.

(2) $\boldsymbol{J}_4=\begin{bmatrix}\lambda_1 & 1\\0 & \lambda_1\end{bmatrix}$.

It is known from linear algebra that $\boldsymbol{A}$ can be transformed into its Jordan canonical

---

[1] stable critical node 稳定临界结点

[2] unstable critical node 不稳定临界结点

form by a nonsingular linear transformation $\boldsymbol{T}_4$:

$$\boldsymbol{T}_4^{-1}\boldsymbol{A}\boldsymbol{T}_4=\boldsymbol{J}_4$$

and $\begin{bmatrix} x \\ y \end{bmatrix}=\boldsymbol{T}_4\begin{bmatrix} \xi \\ \eta \end{bmatrix}$ makes (7.2.5) be

$$\frac{\mathrm{d}}{\mathrm{d}t}\begin{bmatrix} \xi \\ \eta \end{bmatrix}=\boldsymbol{J}_4\begin{bmatrix} \xi \\ \eta \end{bmatrix}$$

with general solution

$$\xi=(C_1+C_2t)\mathrm{e}^{\lambda_1 t},\ \eta=C_2\mathrm{e}^{\lambda_1 t}$$

that is,

$$\begin{bmatrix} \xi \\ \eta \end{bmatrix}=\left(C_1\begin{bmatrix} 1 \\ 0 \end{bmatrix}+C_2\begin{bmatrix} t \\ 1 \end{bmatrix}\right)\mathrm{e}^{\lambda_1 t} \tag{7.2.14}$$

If $C_1=C_2=0$, the trajectory is the singularity $O(0,0)$.

If $C_1\neq 0=C_2$, the orbit coincides with the ξ axis, but with the exception of point O.

If $C_2\neq 0$, eliminating t from (7.2.14),

$$\xi=\left(\frac{C_1}{C_2}+\frac{1}{\lambda_1}\ln\frac{\eta}{C_2}\right)\eta \tag{7.2.14'}$$

So it is known that the trajectory is tangent to the ξ axis at point O. $\lambda_1<0$ gives that all orbits tend to singularity O as $t\to+\infty$, which is called a stable degenerate node[1] (Figure 7.2.9). All orbits tend to O as $t\to-\infty$ when $\lambda_1>0$, this singularity O is called an unstable degenerate node[2] (Figure 7.2.10).

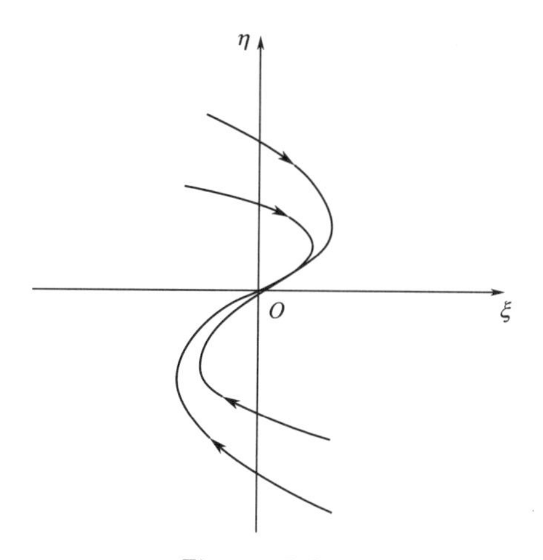

Figure 7.2.9.

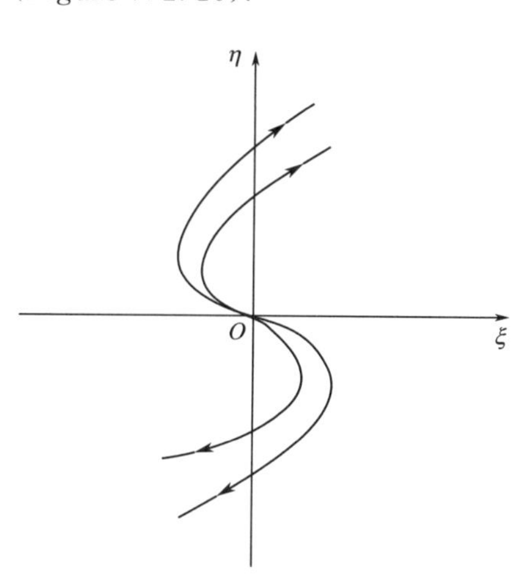

Figure 7.2.10.

---

[1] stable degenerate node 稳定退化结点

[2] unstable degenerate node 不稳定退化结点

The characteristic of the stable (unstable) degenerate node is that it has a neighborhood in which all orbits except the singularity tend to be the singularity as $t \to +\infty$ ($t \to -\infty$) and follow a unique pair of opposite directions respectively.

【Example 7. 2. 3】 Determine the singularity type of the system

$$\begin{cases} \dfrac{dx}{dt} = -3x - y \\ \dfrac{dy}{dt} = 4x + y \end{cases}$$

Solution. The characteristic equation of coefficient matrix $\boldsymbol{A} = \begin{bmatrix} -3 & -1 \\ 4 & 1 \end{bmatrix}$ is

$$\begin{vmatrix} -3-\lambda & -1 \\ 4 & 1-\lambda \end{vmatrix} = \lambda^2 + 2\lambda + 1 = 0$$

The characteristic root $\lambda_1 = -1$ is double real, but coefficients b and c are not all zero, so singularity O is a stable degenerate node.

From the above analysis, combined with the coefficients of the characteristic equation (7. 2. 7) to distinguish the different cases of characteristic roots, the following theorems can be summarized.

Theorem 7. 2. 1. For a two-dimensional linear system (7. 2. 5), let $q \neq 0$ in its characteristic equation (7. 2. 7), that is, neither characteristic roots λ_1 nor λ_2, is zero, then there is the following conclusion about its singularity O:

(1) If $q < 0$, i. e. , unequal λ_1, λ_2 are real number, and $\lambda_1 \lambda_2 < 0$, then O is a saddle point.

(2) If $q > 0$, $p^2 - 4q > 0$, i. e. , unequal λ_1, λ_2 are real number, and $\lambda_1 \lambda_2 > 0$, then O is a stable node when $p > 0$ which gives that $\lambda_1 < 0$, $\lambda_2 < 0$; O is an unstable node when $p < 0$ which gives that $\lambda_1 > 0$, $\lambda_2 > 0$.

(3) If $q > 0$, $p^2 - 4q < 0$, i. e. , λ_1, λ_2 are a pair of conjugate complex numbers: $\lambda_{1,2} = \alpha \pm i\beta$, then O is a center when $p = 0$ ($\alpha = 0$); O is a stable focus when $p > 0$ ($\alpha < 0$); O is an unstable focus when $p < 0$ ($\alpha > 0$).

(4) $q > 0$, $p^2 - 4q = 0$, i. e. , $\lambda_1 = \lambda_2$. If $p > 0$, i. e. , $\lambda_1 = \lambda_2 < 0$, then O is a stable critical node when $b = c = 0$ in (7. 2. 5); O is a stable degenerate node when $b^2 + c^2 \neq 0$ in (7. 2. 5). If $p < 0$, i. e. , $\lambda_1 = \lambda_2 > 0$, then O is an unstable critical node when $b = c = 0$ in (7. 2. 5); O is an unstable degenerate node when $b^2 + c^2 \neq 0$ in (7. 2. 5).

It is convenient to judge the type of singularity by the parameters p and q. A schematic diagram of different types of singularities corresponding to different regions on the parameter (p, q) plane is shown in Figure 7. 2. 11.

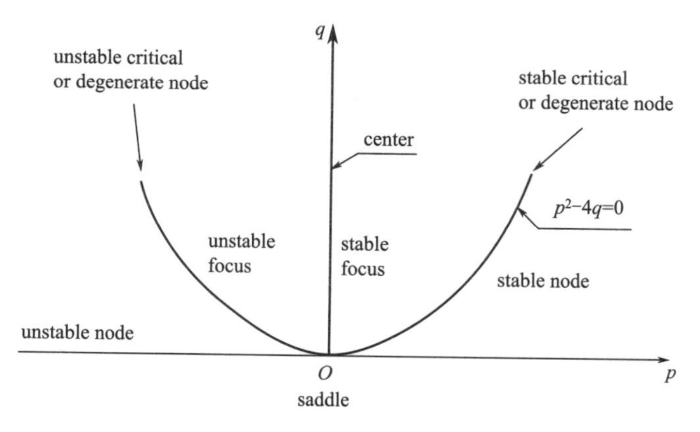

Figure 7.2.11.

7.2.2 Distribution of orbits of two-dimensional nonlinear systems in the neighborhood of singularities

In this part, the distribution of orbits of nonlinear systems (7.2.1)' in the neighborhood of singularity O is studied. We give the following theorem without proof.

Theorem 7.2.2. Let the real part of the root of the characteristic equation (7.2.7) of the first approximation system (7.2.5) for the nonlinear system (7.2.1)' is nonzero, as well as

(1) $R_1(x,y)$ and $R_2(x,y)$ in (7.2.1)' have continuous first-order partial derivatives in a neighborhood of point O;

(2) there exists $\delta>0$ such that

$$\lim_{r\to 0}\frac{R_i(x,y)}{r^{1+\delta}}=0\ (i=1,\ 2),\ r=\sqrt{x^2+y^2} \tag{7.2.15}$$

Then singularities O in (7.2.1)' and (7.2.5) have the same characteristics in their neighborhood, that is, their singularities O are of the same type and are both stable or unstable.

Now consider the case that the singularity is not the origin. Suppose that (x_0,y_0) is a singularity of (7.2.1), that is,

$$X(x_0,y_0)=0,\ Y(x_0,y_0)=0,$$

and the first order Taylor expansion of $X(x,y)$ and $Y(x,y)$ at point (x_0,y_0) is

$$X(x,y)=X(x_0,y_0)+\left.\frac{\partial X}{\partial x}\right|_{(x_0,y_0)}(x-x_0)+\left.\frac{\partial X}{\partial y}\right|_{(x_0,y_0)}(y-y_0)+\widetilde{R}_1(x,y)$$

$$Y(x,y)=Y(x_0,y_0)+\left.\frac{\partial Y}{\partial x}\right|_{(x_0,y_0)}(x-x_0)+\left.\frac{\partial Y}{\partial y}\right|_{(x_0,y_0)}(y-y_0)+\widetilde{R}_2(x,y)$$

Denote

$$\begin{bmatrix} a & b \\ c & d \end{bmatrix} = \begin{bmatrix} \frac{\partial X}{\partial x} & \frac{\partial X}{\partial y} \\ \frac{\partial Y}{\partial x} & \frac{\partial Y}{\partial x} \end{bmatrix}_{(x_0, y_0)}$$

using the coordinate translation $x_1 = x - x_0$, $y_1 = y - y_0$, and

$$\begin{cases} R_1(x_1, y_1) = \widetilde{R}_1(x_1 + x_0, y_1 + y_0) \\ R_2(x_1, y_1) = \widetilde{R}_2(x_1 + x_0, y_1 + y_0) \end{cases}$$

therefore, the system (7.2.1) becomes

$$\begin{cases} \frac{dx_1}{dt} = ax_1 + by_1 + R_1(x_1, y_1) \\ \frac{dy_1}{dt} = cx_1 + dy_1 + R_2(x_1, y_1) \end{cases} \tag{7.2.16}$$

where

$$\lim_{r \to 0} \frac{R_i(x_1, y_1)}{r_1} = \lim_{(x-x_0)^2+(y-y_0)^2 \to 0} \frac{\widetilde{R}_i(x, y)}{\sqrt{(x-x_0)^2+(y-y_0)^2}}$$
$$= 0 \ (i = 1, 2) \tag{7.2.17}$$

Next, we can apply Theorem 7.2.2 on singularity $(x, y) = (0, 0)$.

【Example 7.2.4】 Find the singularities of the system

$$\begin{cases} \frac{dx}{dt} = 2 + y - x^2 \\ \frac{dy}{dt} = 2x(x - y) \end{cases}$$

and determine their types.

Solution. It is easy to find singularities $M_1(0, -2)$, $M_2(2, 2)$, $M_3(-1, -1)$ from equations

$$2 + y - x^2 = 0, \ 2x(x - y) = 0$$

$$\begin{bmatrix} \frac{\partial X}{\partial x} & \frac{\partial X}{\partial y} \\ \frac{\partial Y}{\partial x} & \frac{\partial Y}{\partial x} \end{bmatrix} = \begin{bmatrix} -2x & 1 \\ 4x - 2y & -2x \end{bmatrix}$$

gives matrices $\boldsymbol{A}_1$, $\boldsymbol{A}_2$ and $\boldsymbol{A}_3$ corresponding to M_1, M_2 and M_3. In fact,

$$\boldsymbol{A}_1 = \begin{bmatrix} 0 & 1 \\ 4 & 0 \end{bmatrix}$$

and the characteristic equation is $\lambda^2 - 4 = 0$, so M_1 is a saddle;

$$\boldsymbol{A}_2 = \begin{bmatrix} -4 & 1 \\ 4 & -4 \end{bmatrix}$$

and the characteristic equation is $(\lambda+2)(\lambda+6)=0$, so M_2 is a stable node;

$$\boldsymbol{A}_3=\begin{bmatrix} 2 & 1 \\ -2 & 2 \end{bmatrix}$$

and the characteristic equation is $\lambda^2-4\lambda+6=0$, so M_3 is an unstable focus.

7.3 Limit cycle

It is known from Section 7.1 that the closed orbit corresponds to the periodic solution. The periodic solution is very important, so it is very meaningful to study the closed orbit.

For the closed orbit $x^2+y^2=r_0^2$ (where r_0 is a positive number) of Example 7.1.1 in Section 7.1, there is always a neighborhood of this closed orbit, at each point in which, a closed orbit passes through. Now give an example with different characteristics.

【Example 7.3.1】 Study the closed orbit of the system

$$\begin{cases} \dfrac{dx}{dt}=-y+x(1-x^2-y^2) \\ \dfrac{dy}{dt}=x+y(1-x^2-y^2) \end{cases} \tag{7.3.1}$$

Solution. Introducing polar coordinates

$$x=r\cos\theta,\ y=r\sin\theta$$

system (7.3.1) can be turned into

$$\frac{d\theta}{dt}=1,\ \frac{dr}{dt}=r(1-r^2) \tag{7.3.2}$$

The solution is

$$\begin{cases} \theta=t+\theta_0 \\ r=\dfrac{r_0}{\sqrt{(1-r_0^2)e^{-2t}+r_0^2}} \end{cases} \tag{7.3.3}$$

where $r_0=r(0)$, $\theta_0=\theta(0)$ are the initial values for r, θ.

When $r_0=1$, the corresponding solution is

$$\begin{cases} \theta=t+\theta_0 \\ r=r_0=1 \end{cases} \tag{7.3.4}$$

This is a closed orbit. If $r_0\neq 1$, as can be seen from (7.3.3), whether $r_0>1$ or $0<r_0<1$, the corresponding solution is not a closed orbit, but a spiral rotating infi-

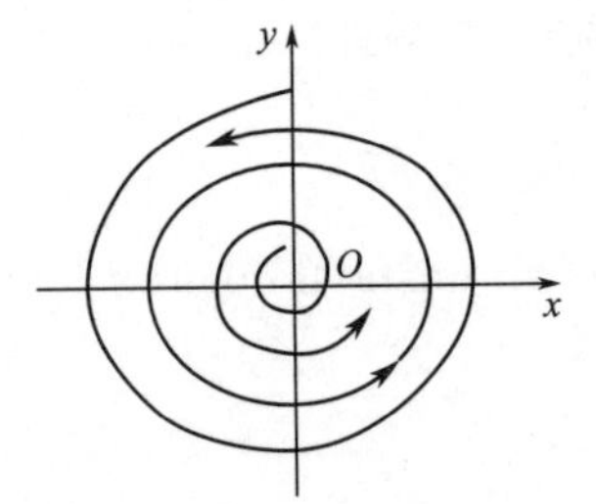

Figure 7.3.1.

nitely around the closed orbit $r=1$. As t increases, it rotates counterclockwise. While $t\to\infty$, $r\to 1$ (Figure 7.3.1).

The closed orbit $r=1$ in Example 7.3.1 differs from that of Example 7.1.1 in Section 7.1: the former has a neighborhood in which every point (except for the points on $r=1$) passes through a non-closed orbit. They rotate infinitely around the closed orbit $r=1$.

Definition 7.3.1. Let L be a closed orbit of a two-dimensional autonomous system. If there is an outer (inner) side neighborhood of L, every point in that passes through a non closed orbit, and they rotate infinitely around the closed orbit L. When $t\to+\infty$, the distance between the point on the non-closed orbit and L tends to zero. The closed orbit L is called the outer (inner) side stable limit cycle[1].

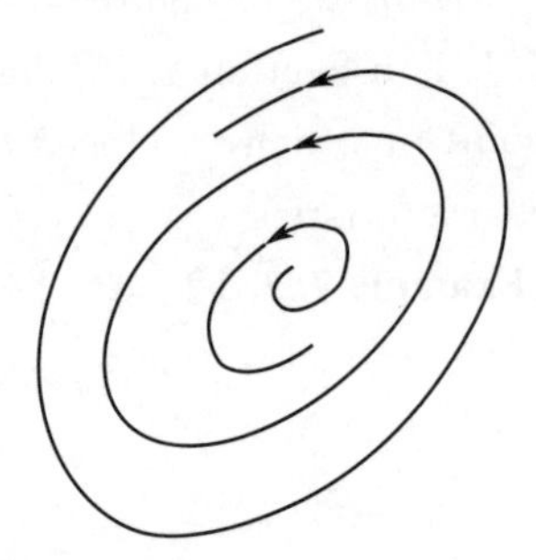

Figure 7.3.2.

If $t\to+\infty$ in the above definition is changed to $t\to-\infty$, the corresponding closed orbit L is called the outer (inner) side unstable limit cycle[2].

The stable (unstable) limit cycle on both sides is called stable (unstable) limit cycle[3], and the stable on one side and unstable limit cycle on the other side are called semi-stable limit cycle[4].

Figures 7.3.2, 7.3.3 and 7.3.4 are schematic diagrams of stable limit cycle, unstable limit cycle and semi-stable limit cycle, respectively.

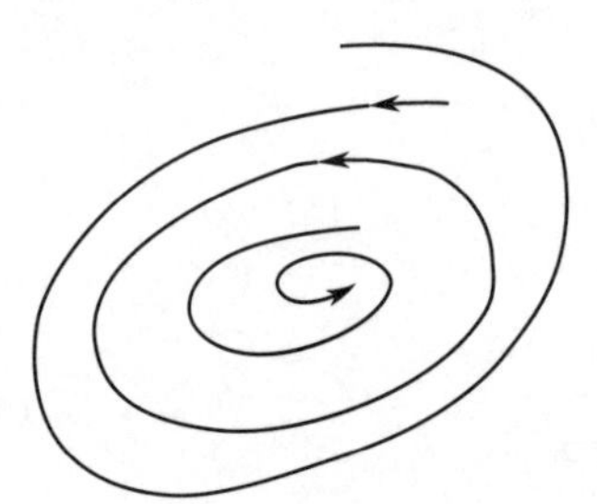

Figure 7.3.3.

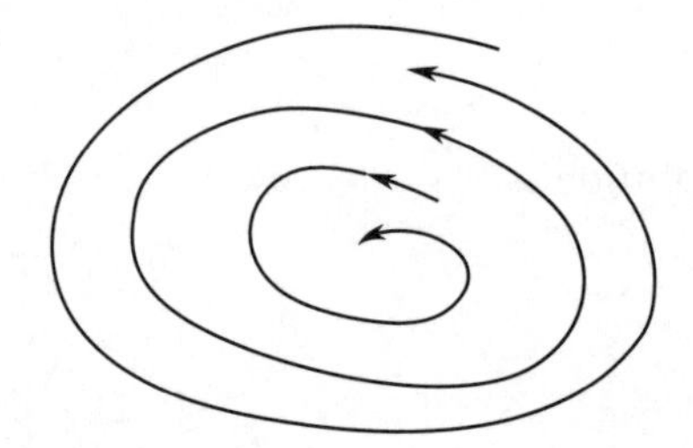

Figure 7.3.4.

---

[1] outer (inner) side stable limit cycle 外（内）侧稳定极限环

[2] outer (inner) side unstable limit cycle 外（内）侧不稳定极限环

[3] stable (unstable) limit cycle 稳定（不稳定）极限环

[4] semi-stable limit cycle 半稳定极限环

Stable limit cycle is of great significance in practical problems. Because the initial value of the practical problem is not necessarily located on the trajectory L of a desired periodic solution. If L is a stable limit cycle, then there exists one of its neighborhoods, as long as the point (x_0, y_0) is in this neighborhood, the trajectory of the solution $x=x(t)$, $y=y(t)$ corresponding to $x(0)=x_0$, $y(0)=y_0$ as the initial value enters any small neighborhood of L when t is sufficiently large.

That is, when t is sufficiently large, the solution $x=x(t)$, $y=y(t)$ is actually represented by the periodic solution corresponding to L. In other words, a stable limit cycle can withstand a small disturbance of the initial value. As for the unstable limit cycle, it is different. As long as the initial value deviates from the unstable limitcycle L, the solution corresponding to the initial value will be far away from L and not return to the neighborhood of L when t increases.

7.4 Lyapunov stability

Stability theory was founded by Russian mathematician Lyapunov in 1880s. It is widely used in natural science and engineering technology such as automatic control, aerospace technology, ecological biology, biochemical reaction and so on. Its concept and theory are developing rapidly. Next, the preliminary theory of Lyapunov stability is introduced in two sections.

7.4.1 Stability

Consideration system

$$\frac{\mathrm{d}\boldsymbol{y}}{\mathrm{d}t}=\boldsymbol{\psi}(t, \boldsymbol{y}) \tag{7.4.1}$$

where $\boldsymbol{y}$ and $\boldsymbol{\psi}$ are n-dimensional vector functions, $(t, \boldsymbol{y}) \in D \subset \mathbf{R}^{n+1}$. Let $\boldsymbol{\psi}(t, \boldsymbol{y})$ be continuous in the open region D and satisfy the local Lipschitz condition for $\boldsymbol{y}$ in D. Suppose that $\boldsymbol{y}=\boldsymbol{\varphi}(t)$ be the solution of (7.4.1), which exists on $T_0 \leqslant t \leqslant T$. $(t, \boldsymbol{\varphi}(t)) \in D$ as $t \in [t_0, T]$. From the theorem of continuous dependence of the solution of the system on the initial value, it is easy to deduce the following facts: For any given $\varepsilon>0$ and $t_0 \in [T_0, T]$, there is $\delta=\delta(\varepsilon, t_0)>0$, as long as $\boldsymbol{y}_0$ satisfies

$$\|\boldsymbol{y}_0-\boldsymbol{\varphi}(t_0)\|<\delta$$

then the solution $\boldsymbol{y}=\boldsymbol{y}(t, t_0, \boldsymbol{y}_0)$ satisfying the initial value $\boldsymbol{y}(t_0)=\boldsymbol{y}_0$ also exists on the interval $[T_0, T]$, and for all $t \in [T_0, T]$,

$$\|\boldsymbol{y}(t, t_0, \boldsymbol{y}_0)-\boldsymbol{\varphi}(t)\|<\varepsilon$$

that is, on the finite interval $[T_0, T]$, as long as the initial value changes small enough, The change of the solution can be less than any small amount specified in advance. In the case of $t\in[T_0,T]$, it is uniformly to get

$$\lim_{\boldsymbol{y}_0\to\boldsymbol{\varphi}(t_0)}\boldsymbol{y}(t,t_0,\boldsymbol{y}_0)=\boldsymbol{\varphi}(t)$$

Now extend the problem to the infinite interval $T_0\leqslant t<+\infty$, what will happen? Look at the example.

【Example 7.4.1】 Consider the initial value problem on $0\leqslant t<+\infty$

$$\begin{cases}\dfrac{dx}{dt}=\dfrac{\lambda x}{t+1}-\dfrac{1}{(t+1)^2}\ (\lambda\neq-1)\\ x(t_0)=x_0,\ t_0\in[0,+\infty)\end{cases}\tag{7.4.2}$$

whose solution is

$$x=\left(\frac{t+1}{t_0+1}\right)^{\lambda}x_0+\frac{1}{\lambda+1}\left[\frac{1}{t+1}-\frac{(t+1)^{\lambda}}{(t_0+1)^{\lambda+1}}\right]\tag{7.4.3}$$

$x_0=0$ gives that

$$x=\frac{1}{\lambda+1}\left[\frac{1}{t+1}-\frac{(t+1)^{\lambda}}{(t_0+1)^{\lambda+1}}\right]\tag{7.4.4}$$

denoted by $x=\varphi(t)$. Examine the absolute of the difference $\Delta(t,t_0,x_0)$ between solutions (7.4.3) and (7.4.4), that is,

$$|\Delta(t,t_0,x_0)|=|x(t)-\varphi(t)|=\left(\frac{t+1}{t_0+1}\right)^{\lambda}|x_0|\tag{7.4.5}$$

Discuss in two cases:

(1) $\lambda\leqslant0$. (7.4.5) tells us that for any $t\in[t_0,+\infty)$,

$$|\Delta(t,t_0,x_0)|<\frac{|x_0|}{(t_0+1)^{\lambda}}$$

So for any given $\varepsilon>0$, set

$$\delta=\varepsilon(t_0+1)^{\lambda}\tag{7.4.6}$$

and when $|x_0|<\delta$, we have

$$|\Delta(t,t_0,x_0)|<\varepsilon,t\in[t_0,+\infty)\tag{7.4.7}$$

that is, for $t\in[t_0,+\infty)$, there uniformly exists

$$\lim_{x_0\to0}(x(t)-\varphi(t))=0\tag{7.4.8}$$

(2) $\lambda>0$. Given $\varepsilon_0>0$, no matter how small $\delta>0$ is, if δ and x_0 satisfy $|x_0|<\delta$, it is easy to know by (7.4.5): when

$$t\geqslant(t_0+1)\left(\frac{\varepsilon_0}{|x_0|}\right)^{\frac{1}{\lambda}}-1\triangleq T_1\tag{7.4.9}$$

$$|\Delta(t,t_0,x_0)|\geqslant\varepsilon_0\tag{7.4.10}$$

That is to say, in this case, on the infinite interval $[t_0,+\infty)$, the absolute of the difference between the two solutions does not decrease uniformly because the absolute of the difference between their corresponding initial values becomes smaller (see Figure 7.4.1).

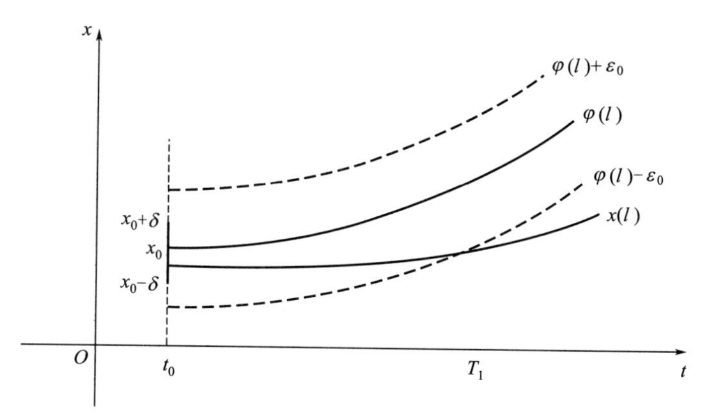

Figure 7.4.1

It is also pointed out that in the case of $\lambda \leqslant 0$, divided into two cases: $\lambda < 0$ and $\lambda = 0$. When $\lambda < 0$, as seen from (7.4.5), $\delta_1 = (t_0 + 1)^{\lambda} > 0$ can be set, when $0 < |x_0| < \delta_1$ (see Figure 7.4.2),

$$|x(t) - \varphi(t)| = \left(\frac{t+1}{t_0+1}\right)^{\lambda} |x_0| < (t+1)^{\lambda}$$

So

$$\lim_{t \to +\infty} (x(t) - \varphi(t)) = 0$$

For $\lambda = 0$, $|x(t) - \varphi(t)| = |x_0|$ can only be obtained from (7.4.5), thus

$$\lim_{t \to +\infty} (x(t) - \varphi(t)) \neq 0$$

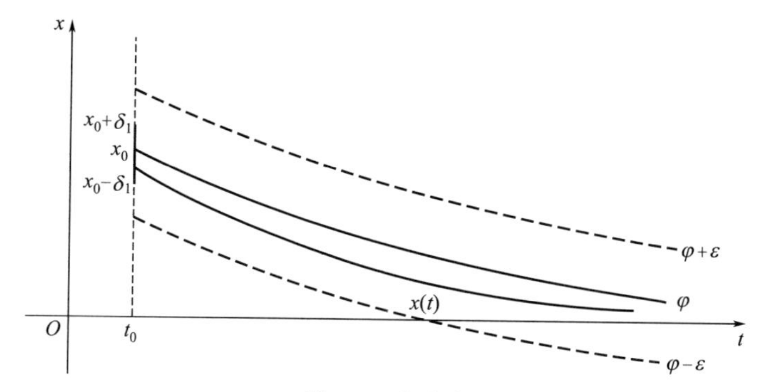

Figure 7.4.2

It can be seen that when $\lambda = 0$, $\lambda < 0$ and $\lambda > 0$, the corresponding solutions in above example show very different properties on the infinite interval $[T_0, +\infty)$.

We consider the solution $\boldsymbol{y} = \boldsymbol{\varphi}(t)$ of the system (7.4.1) and let it exist on the interval $0 \leqslant t < +\infty$. Lyapunov introduced the following definition.

Definition 7.4.1. For any given $\varepsilon>0$ and $t_0\geqslant 0$, there be $\delta=\delta(\varepsilon,t_0)>0$ as long as y_0 satisfies

$$\|\boldsymbol{y}_0-\boldsymbol{\varphi}(t_0)\|<\delta$$

the solution $\boldsymbol{y}=\boldsymbol{y}(t,t_0,\boldsymbol{y}_0)$ of the system (7.4.1) with the initial value $\boldsymbol{y}(t_0)=\boldsymbol{y}_0$ always exists on the interval $[t_0,+\infty)$, and for all $t\in[t_0,+\infty)$, there is

$$\|\boldsymbol{y}(t,t_0,\boldsymbol{y}_0)-\boldsymbol{\varphi}(t)\|<\varepsilon \tag{7.4.11}$$

That is, for $t_0\leqslant t<+\infty$, it holds uniformly that

$$\lim_{\boldsymbol{y}_0\to\boldsymbol{\varphi}(t_0)}[\boldsymbol{y}(t,t_0,\boldsymbol{y}_0)-\boldsymbol{\varphi}(t)]=\boldsymbol{0} \tag{7.4.12}$$

then it is said that the solution $\boldsymbol{y}=\boldsymbol{\varphi}(t)$ of (7.4.1) is stable. [1]

If the solution $\boldsymbol{y}=\boldsymbol{\varphi}(t)$ is not only stable, but also for any given $t_0\geqslant 0$, there is $\delta_1=\delta_1(t_0)>0$, as long as $\boldsymbol{y}_0$ satisfies

$$\|\boldsymbol{y}_0-\boldsymbol{\varphi}(t_0)\|<\delta_1$$

the solution $\boldsymbol{y}=\boldsymbol{y}(t,t_0,\boldsymbol{y}_0)$ of the system (7.4.1) with the initial value $\boldsymbol{y}(t_0)=\boldsymbol{y}_0$ gives

$$\lim_{t\to+\infty}[\boldsymbol{y}(t,t_0,\boldsymbol{y}_0)-\boldsymbol{\varphi}(t)]=\boldsymbol{0} \tag{7.4.13}$$

then it is said that the solution $\boldsymbol{y}=\boldsymbol{\varphi}(t)$ of (7.4.1) is asymptotically stable. [2]

For some $\varepsilon_0>0$ and $t_0\geqslant 0$, no matter how small $\delta>0$ is, there is always $\boldsymbol{y}_0$ such that $\|\boldsymbol{y}_0-\boldsymbol{\varphi}(t_0)\|<\delta$, but (7.4.1) has a solution $\boldsymbol{y}(t,t_0,\boldsymbol{y}_0)$ with the initial value $\boldsymbol{y}(t_0)=\boldsymbol{y}_0$ satisfying as $t=t_1(\geqslant t_0)$

$$\|\boldsymbol{y}(t_1,t_0,y_0)-\boldsymbol{\varphi}(t_1)\|\geqslant\varepsilon_0 \tag{7.4.14}$$

then the solution $\boldsymbol{y}=\boldsymbol{\varphi}(t)$ of (7.4.1) is unstable. [3]

According to this definition, in Example 7.4.1, the solution (7.4.4) is stable, asymptotically stable and unstable when $\lambda=0$, $\lambda<0$ and $\lambda>0$.

The following part is devoted to Lyapunov stability theory. For that, the solution $\boldsymbol{y}=\boldsymbol{\varphi}(t)$ in system (7.4.1) is transformed into zero solution, so that it is more standardized in narration and argumentation.

A variable transformation is made for the system (7.4.1)

$$\boldsymbol{x}=\boldsymbol{y}-\boldsymbol{\varphi}(t) \tag{7.4.15}$$

Under the change (7.4.15), the system (7.4.1) becomes

$$\frac{\mathrm{d}\boldsymbol{x}}{\mathrm{d}t}=\frac{\mathrm{d}\boldsymbol{y}}{\mathrm{d}t}-\frac{\mathrm{d}\boldsymbol{\varphi}(t)}{\mathrm{d}t}=\boldsymbol{\psi}(t,\boldsymbol{x}+\boldsymbol{\varphi}(t))-\boldsymbol{\psi}(t,\boldsymbol{\varphi}(t))$$

Denote the right side of the above formula as $\boldsymbol{f}(t,\boldsymbol{x})$, so (7.4.1) becomes

---

[1] stable 稳定的

[2] asymptotically stable 渐近稳定的

[3] unstable 不稳定的

$$\frac{\mathrm{d}\boldsymbol{x}}{\mathrm{d}t}=\boldsymbol{f}(t,\boldsymbol{x}) \tag{7.4.16}$$

where

$$\boldsymbol{f}(t,\boldsymbol{0})=\boldsymbol{0} \tag{7.4.17}$$

The solution $\boldsymbol{y}=\boldsymbol{\varphi}(t)$ of the system (7.4.1) is transformed into the zero solution $\boldsymbol{x}=\boldsymbol{0}$ of the system (7.4.16). We just need to discuss the stability of the system (7.4.16) about $\boldsymbol{x}=\boldsymbol{0}$ to get the information of $\boldsymbol{y}=\boldsymbol{\varphi}(t)$ of the system (7.4.1).

Now, suppose that

$$\boldsymbol{f}(\boldsymbol{0})=\boldsymbol{0} \tag{7.4.18}$$

and we investigate the stability of zero solution $\boldsymbol{x}=\boldsymbol{0}$ of autonomous system

$$\frac{\mathrm{d}\boldsymbol{x}}{\mathrm{d}t}=\boldsymbol{f}(\boldsymbol{x}) \tag{7.4.19}$$

In addition to satisfying the condition (7.4.18), $\boldsymbol{f}(\boldsymbol{x})$ has a first-order continuous partial derivative on region

$$G=\{\boldsymbol{x} \mid \|\boldsymbol{x}\|<H\} \tag{7.4.20}$$

where H is a positive number.

7.4.2 First approximation theory

Under these assumptions mentioned at the end of the preceding paragraph, when the initial value $\boldsymbol{x}(t_0)=\boldsymbol{x}_0 \in G$, the solution of (7.4.19) exists and is unique, and $\boldsymbol{f}(\boldsymbol{x})$ can be expanded according to Taylor's formula on region G

$$\boldsymbol{f}(\boldsymbol{x})=\boldsymbol{A}\boldsymbol{x}+\boldsymbol{R}(\boldsymbol{x}) \tag{7.4.21}$$

where

$$\boldsymbol{A}=\begin{bmatrix} \frac{\partial f_1}{\partial x_1} & \frac{\partial f_1}{\partial x_2} & \cdots & \frac{\partial f_1}{\partial x_n} \\ \frac{\partial f_2}{\partial x_1} & \frac{\partial f_2}{\partial x_2} & \cdots & \frac{\partial f_2}{\partial x_n} \\ \vdots & \vdots & & \vdots \\ \frac{\partial f_n}{\partial x_1} & \frac{\partial f_n}{\partial x_2} & \cdots & \frac{\partial f_n}{\partial x_n} \end{bmatrix}_{\boldsymbol{x}=\boldsymbol{0}} \tag{7.4.22}$$

is an n-order constant matrix, $\boldsymbol{f}=(f_1, f_2,\cdots,f_n)^{\mathrm{T}}$, $\boldsymbol{x}=(x_1,x_2,\cdots,x_n)^{\mathrm{T}}$, $\boldsymbol{R}(\boldsymbol{x})$ is a vector function with

$$\lim_{\|\boldsymbol{x}\|\to 0}\frac{\|\boldsymbol{R}(\boldsymbol{x})\|}{\|\boldsymbol{x}\|}=0 \tag{7.4.23}$$

Substituting (7.4.21) into (7.4.19) shows that

$$\frac{\mathrm{d}\boldsymbol{x}}{\mathrm{d}t}=\boldsymbol{A}\boldsymbol{x}+\boldsymbol{R}(\boldsymbol{x}) \tag{7.4.19$'$}$$

Then the system

$$\frac{\mathrm{d}\boldsymbol{x}}{\mathrm{d}t}=\boldsymbol{A}\boldsymbol{x} \tag{7.4.24}$$

is a first approximation system of (4. 19) (i. e. (4. 19)'), which is a system of linear homogeneous equations with constant coefficients.

It is natural to think that from the stability of the zero solution $\boldsymbol{x}=\boldsymbol{0}$ of the first approximation system (7. 4. 24) of (7. 4. 19), can we obtain the information of the stability of $\boldsymbol{x}=\boldsymbol{0}$ of (7. 4. 19)? Now, we first study (7. 4. 24), and there is the following theorem about its stability.

Theorem 7. 4. 1.

(1) If the real parts of all characteristic roots of $\boldsymbol{A}$ are negative, then the zero solution of (7. 4. 24) is asymptotically stable;

(2) If at least a real part of characteristic roots of $\boldsymbol{A}$ is positive, then the zero solution of (7. 4. 24) is unstable;

(3) If the real parts of the characteristic root of $\boldsymbol{A}$ are not positive and have zero real parts, then the zero solution of (7. 4. 24) may be unstable or stable. But it is not be asymptotically stable.

Remark 7. 4. 1. It can be seen from Theorem 7. 4. 1 that the zero solution $\boldsymbol{x}=\boldsymbol{0}$ of (7. 4. 24) is asymptotically stable if and only if the real parts of all characteristic roots of $\boldsymbol{A}$ are negative.

Remark 7. 4. 2. Let the real parts of the characteristic root of $\boldsymbol{A}$ be not positive, but there are zero real parts, then when the number of linearly independent eigenvectors corresponding to each characteristic root with zero real part is equal to the multiplicity of the characteristic root, $\boldsymbol{x}=\boldsymbol{0}$ is stable and not asymptotically stable; when the number of linearly independent eigenvectors corresponding to at least one characteristic root with zero real part is less than the its multiplicity, $\boldsymbol{x}=\boldsymbol{0}$ is unstable.

【Example 7. 4. 2】 Give the stability of zero solution of the following system:

$$\begin{cases}\dfrac{\mathrm{d}x_1}{\mathrm{d}t}=-10x_1+5x_2+x_3+2x_4\\[2mm] \dfrac{\mathrm{d}x_2}{\mathrm{d}t}=12x_1+6x_2-2x_3-4x_4\\[2mm] \dfrac{\mathrm{d}x_3}{\mathrm{d}t}=-16x_1-8x_2-4x_3-8x_4\\[2mm] \dfrac{\mathrm{d}x_4}{\mathrm{d}t}=6x_1+3x_2+x_3+2x_4\end{cases}$$

Solution. The characteristic equation is $\lambda^2(\lambda+2)(\lambda+4)=0$. $\lambda=0$ is a double root, and the real parts of the other two roots are negative. It is known from Remark 7. 4. 1, the zero solution is not asymptotically stable. In order to study whether the zero solution is stable or unstable, only the number of linearly independent eigenvectors corresponding to the characteristic root $\lambda=0$ is examined. Next, we check the rank of matrix

$$\begin{bmatrix} -10-0 & 5 & 1 & 2 \\ 12 & 6-0 & -2 & -4 \\ -16 & -8 & -4-0 & -8 \\ 6 & 3 & 1 & 2-0 \end{bmatrix} \tag{7.4.25}$$

The first and fourth columns of the matrix can be reduced to zero by column transformation

$$\begin{bmatrix} 0 & 5 & 1 & 0 \\ 0 & 6 & -2 & 0 \\ 0 & -8 & -4 & 0 \\ 0 & 3 & 1 & 0 \end{bmatrix}$$

Therefore, it is known that the rank of (7.4.25) is 2, and for the characteristic root $\lambda=0$, $4-2=2$ linearly independent eigenvectors can be obtained. The number of linearly independent eigenvectors is equal to the multiplicity of the eigenvalues $\lambda=0$. As shown in Remark 7.4.1, the zero solution of the system is stable but not asymptotically stable.

【Example 7.4.3】 Give the stability of zero solution of the following system:

$$\begin{cases} \dfrac{dx}{dt}=-9x+2y+4z \\ \dfrac{dy}{dt}=32x-8y-15z \\ \dfrac{dz}{dt}=-36x+8y+16z \end{cases}$$

Solution. The characteristic equation is $-\lambda^2(\lambda+1)=0$ with roots $\lambda_1=-1$, $\lambda_2=0$ (double), and check the rank of matrix

$$\begin{bmatrix} -9-0 & 2 & 4 \\ 32 & -8-0 & -15 \\ -36 & 8 & 16-0 \end{bmatrix} \tag{7.4.26}$$

The rank is 2 by elementary transformation. Therefore, for the characteristic root $\lambda_2=0$, $3-2=1$ (less than the multiplicity) linearly independent eigenvectors can be obtained. As shown in Remark 7.4.1, the zero solution of the system is unstable.

The zero solution of linear system has been discussed. Now we give the stability of zero solution of nonlinear system (7.4.19)' by the following theorem:

Theorem 7.4.2. $\boldsymbol{R}(\boldsymbol{x})$ in the system (7.4.19)' satisfy the condition (7.4.23).

(1) If the real parts of all characteristic roots of $\boldsymbol{A}$ are negative, then the zero solu-

tion of (7.4.19)' is asymptotically stable;

(2) If at least one real part of the characteristic roots of $\boldsymbol{A}$ is positive, then the zero solution of (7.4.19)' is unstable.

Remark 7.4.3. As can be seen from Theorem 7.4.2, when $\boldsymbol{R}(\boldsymbol{x})$ in (7.4.19)' satisfies the condition (7.4.23), if all the real parts of the characteristic roots of $\boldsymbol{A}$ are negative, or the real part of at least one is positive, then the zero solution of (7.4.19)' and its first approximation system (7.4.24) is both asymptotically stable or unstable.

【Example 7.4.4】 Research the stability of zero solution $x=y=z=0$ of the system

$$\begin{cases} \dfrac{dx}{dt}=y-z-2\sin x \\ \dfrac{dy}{dt}=x-2y+(\sin y+z^2)e^x \\ \dfrac{dz}{dt}=x+y-\dfrac{z}{1-z}-\sin y \end{cases}$$

Solution. The first approximate system of the original system is obtained by expanding $\sin x$, $\sin y$, e^x, $\dfrac{z}{1-z}$ in the neighborhood of $x=y=z=0$ into Taylor's formula and omitting the higher order term:

$$\begin{cases} \dfrac{dx}{dt}=-2x+y-z \\ \dfrac{dy}{dt}=x-y \\ \dfrac{dz}{dt}=x-z \end{cases}$$

The characteristic equation of coefficient matrix $\boldsymbol{A}$ is

$$\det(\boldsymbol{A}-\lambda\boldsymbol{I})=-(\lambda+2)(\lambda+1)^2=0$$

All of its roots have negative real parts, so the zero solution of the original system is asymptotically stable.

【Example 7.4.5】 Find the stability of zero solution of the system

$$\begin{cases} \dfrac{dx}{dt}=e^{x+y}+z-1 \\ \dfrac{dy}{dt}=2x+y+\sin z \\ \dfrac{dz}{dt}=-8x-5y-3z+xy^2 \end{cases}$$

Solution. The first approximate system of the original system is

$$\begin{cases}\dfrac{dx}{dt}=x+y+z \\ \dfrac{dy}{dt}=2x+y-z \\ \dfrac{dz}{dt}=-8x-5y-3z\end{cases}$$

Its characteristic equation

$$\lambda^3+\lambda^2-4\lambda-4=0$$

It easy to see the eigenvalues are -1, -2, 2. Thus, the zero solution of the original system is unstable.

Remark 7.4.4. Theorem 7.4.2 does not mention the case that "the real parts of the characteristic roots of $\boldsymbol{A}$ are not positive, but there are zero real parts". In this case, the zero solution of the nonlinear system (7.4.19)' may be stable, unstable, or asymptotically stable.

【Example 7.4.6】 Give the stability of the zero solution $x=0$ of the system

$$\frac{dx}{dt}=\alpha x^3 \tag{7.4.27}$$

where α is a constant.

Solution. Through the integral, it is easy to obtain the solution of (7.4.27) satisfying condition $x(0)=x_0$

$$x(t)=\frac{x_0}{\sqrt{1-2\alpha x_0^2 t}} \tag{7.4.28}$$

If $\alpha<0$, then (7.4.28) is defined on interval $[0,+\infty)$. For any given $\varepsilon>0$, take $\delta=\varepsilon$, then when $0<|x_0|<\delta$, for all $t\in[0,+\infty)$, there is

$$|x(t)|\leqslant|x_0|<\delta=\varepsilon$$

Taking $\delta_1=1$, when $0<|x_0|<\delta_1$,

$$\lim_{t\to+\infty}x(t)=0$$

It demonstrates that the zero solution of (7.4.27) is asymptotically stable.

If $\alpha>0$, for the given ε_0, no matter how small $\delta>0$ $(\delta<\varepsilon_0)$, and taking x_0 such that $0<|x_0|<\delta$, when

$$\frac{\varepsilon_0^2-x_0^2}{2\alpha\varepsilon_0^2x_0^2}\leqslant t<\frac{1}{2\alpha x_0^2}$$

it holds that

$$|x(t)|=\frac{|x_0|}{\sqrt{1-2\alpha x_0^2 t}}\geqslant\varepsilon_0$$

This shows that it is impossible to have $|x(t)|\leqslant\varepsilon_0$ on the interval of $0\leqslant t<+\infty$.

That is, the zero solution of (7. 4. 27) is unstable.

The first approximation system of (7. 4. 27) is $\frac{dx}{dt}=0$, and the eigenvalue is zero. From the above analysis, it can be seen that the stability of the zero solution of (7. 4. 27) cannot be determined by that of its first approximation system.

7. 5 Exercises

1. Given the following planar autonomous systems:

(1) $\begin{cases}\frac{dx}{dt}=y,\\ \frac{dy}{dt}=g;\end{cases}$ (2) $\begin{cases}\frac{dx}{dt}=\frac{1}{k},\\ \frac{dy}{dt}=-y.\end{cases}$

Try to find their solutions, then write the trajectory equation, and draw them on the phase plane, where g and k are constants.

2. For nonautonomous system $\frac{dx}{dt}=x$, $\frac{dy}{dt}=y+t$, check whether Lemmas 7. 1. 1, 7. 1. 2 and 7. 1. 4 are valid.

3. Let $x=\varphi(t)$, $y=\psi(t)$ be a nonconstant solution of (7. 1. 3) and there exist two numbers t_0 and T such that $\varphi(t_0+T)=\varphi(t_0)$, $\psi(t_0+T)=\psi(t_0)$, $T\neq 0$. Prove that for all t,

$$\varphi(t+T)=\varphi(t),\ \psi(t+T)=\psi(t)$$

that is, $x=\varphi(t)$, $y=\psi(t)$ is a periodic solution, whose corresponding trajectory is a closed orbit.

4. Determine the type of singularity for each of the following systems:

(1) $\begin{cases}\frac{dx}{dt}=3x+2y,\\ \frac{dy}{dt}=x+4y;\end{cases}$ (2) $\begin{cases}\frac{dx}{dt}=x+y,\\ \frac{dy}{dt}=-5x-3y;\end{cases}$

(3) $\begin{cases}\frac{dx}{dt}=-2x-y,\\ \frac{dy}{dt}=x;\end{cases}$ (4) $\begin{cases}\frac{dx}{dt}=x-y,\\ \frac{dy}{dt}=5x-4y;\end{cases}$

(5) $\begin{cases}\frac{dx}{dt}=3x+y,\\ \frac{dy}{dt}=-x+y;\end{cases}$ (6) $\begin{cases}\frac{dx}{dt}=3x+4y,\\ \frac{dy}{dt}=2x+y;\end{cases}$

(7) $\begin{cases} \dfrac{dx}{dt}=x-y, \\ \dfrac{dy}{dt}=x+y; \end{cases}$ (8) $\begin{cases} \dfrac{dx}{dt}=x, \\ \dfrac{dy}{dt}=y; \end{cases}$

(9) $\begin{cases} \dfrac{dx}{dt}=x+5y, \\ \dfrac{dy}{dt}=-x-y. \end{cases}$

5. Find the singularities of the following systems and determine their types:

(1) $\begin{cases} \dfrac{dx}{dt}=-x-y+1, \\ \dfrac{dy}{dt}=x-y-5; \end{cases}$ (2) $\begin{cases} \dfrac{dx}{dt}=x+y(1+y-x), \\ \dfrac{dy}{dt}=x(2+x). \end{cases}$

6. Introducing $y=\dfrac{dx}{dt}$, the equation $\dfrac{d^2x}{dt^2}+b\dfrac{dx}{dt}+cx=0$ is transformed into a set of equations. Discuss the types of singularities of the system in the light of the various cases of parameters b and $c(c\neq 0)$.

7. Let $q=ad-bc\neq 0$ in system (7.2.5), and prove that the system has only a unique singularity $O(0,0)$.

8. Find the limit cycles of the following systems and explain their stability:

(1) $\dfrac{dx}{dt}=X(x,y)$, $\dfrac{dy}{dt}=Y(x,y)$, where

$$X(x,y)=\begin{cases} y-x(x^2+y^2-1)(x^2+y^2)^{-\frac{1}{2}}, & \text{if } x^2+y^2\neq 0, \\ 0, & \text{if } x^2+y^2=0; \end{cases}$$

$$Y(x,y)=\begin{cases} -x-y(x^2+y^2-1)(x^2+y^2)^{-\frac{1}{2}}, & \text{if } x^2+y^2\neq 0, \\ 0, & \text{if } x^2+y^2=0; \end{cases}$$

(2) $\begin{cases} \dfrac{dx}{dt}=-x+(x-y)(x^2+y^2)^{\frac{1}{2}}, \\ \dfrac{dy}{dt}=-y+(x+y)(x^2+y^2)^{\frac{1}{2}}. \end{cases}$

9. Let $P(x,y)$ and $Q(x,y)$ be continuous in a simply connected region G and have continuous first-order partial derivatives. L is the closed orbit of the system

$$\frac{dx}{dt}=P(x,y),\ \frac{dy}{dt}=Q(x,y)$$

which is completely located in G, and the region surrounded by L is denoted by D. Prove that

$$\iint_D \left(\frac{\partial P}{\partial x}+\frac{\partial Q}{\partial y}\right)\mathrm{d}x\,\mathrm{d}y=0.$$

10. Show that the system

$$\frac{\mathrm{d}x}{\mathrm{d}t}=x+2xy+x^3,\ \frac{\mathrm{d}y}{\mathrm{d}t}=-y^2+x^2y$$

does not have a limit cycle in the whole plane.

11. Let $\boldsymbol{f}(t,\boldsymbol{0})\equiv\boldsymbol{0}$, try to describe the definitions that zero solution of the system (7.4.16) is stable, asymptotically stable and unstable.

12. Let the n-order matrix function $\boldsymbol{A}(t)$ and n-d vector function $\boldsymbol{f}(t)$ are continuous on $[0,+\infty)$. Try to prove that if

$$\frac{\mathrm{d}\boldsymbol{x}}{\mathrm{d}t}=\boldsymbol{A}(t)\boldsymbol{x}+\boldsymbol{f}(t)$$

has a stable (an asymptotically stable, an unstable) solution, then all the solutions of the system are stable (asymptotically stable, unstable).

13. For the question above, show that any solution of

$$\frac{\mathrm{d}\boldsymbol{x}}{\mathrm{d}t}=\boldsymbol{A}(t)\boldsymbol{x}+\boldsymbol{f}(t)$$

is stable (asymptotically stable, unstable) if and only if its corresponding homogeneous system of equations

$$\frac{\mathrm{d}\boldsymbol{x}}{\mathrm{d}t}=\boldsymbol{A}(t)\boldsymbol{x}$$

has stable (asymptotically stable, unstable) zero solution.

14. Suppose that $f(t,x)$ and $f_x(t,x)$ are continuous in $D=\{(t,x)\mid t\geqslant 0,\ |x|<A\}$, together with $f(t,0)\equiv 0$. The zero solution $x=0$ of the equation

$$\frac{\mathrm{d}x}{\mathrm{d}t}=f(t,x) \qquad (*)$$

is stable, and there exist $x_1>0$ and $x_2<0$ so that the solutions $x=\varphi_1(t)$ with $x(0)=x_1$ and $x=\varphi_2(t)$ with $x(0)=x_2$ go to zero as $t\to+\infty$. Show that the zero solution of (*) is asymptotically stable.

15. Give the stability of zero solution in the following systems:

(1) $\begin{cases}\dfrac{\mathrm{d}x}{\mathrm{d}t}=-ky,\\ \dfrac{\mathrm{d}y}{\mathrm{d}t}=kx-2ny;\end{cases}\quad (k>n>0)$

(2) $\begin{cases}\dfrac{\mathrm{d}x}{\mathrm{d}t}=-x-y+2z,\\ \dfrac{\mathrm{d}y}{\mathrm{d}t}=x-y,\\ \dfrac{\mathrm{d}z}{\mathrm{d}t}=x-y;\end{cases}$

(3) $\begin{cases} \dfrac{dx_1}{dt}=-5x_1-5x_2+x_3+x_4, \\ \dfrac{dx_2}{dt}=3x_1+3x_2-x_3-x_4, \\ \dfrac{dx_3}{dt}=-4x_1-4x_2-2x_3-2x_4, \\ \dfrac{dx_4}{dt}=3x_1+3x_2+x_3+x_4; \end{cases}$

(4) $\begin{cases} \dfrac{dx_1}{dt}=-x_1-x_2+x_3+x_1x_2, \\ \dfrac{dx_2}{dt}=x_1-2x_2+2x_3+x_3^3, \\ \dfrac{dx_3}{dt}=x_1+2x_2+x_3+x_1x_3; \end{cases}$

(5) $\begin{cases} \dfrac{dx_1}{dt}=x_1+e^{x_2}-\cos x_2, \\ \dfrac{dx_2}{dt}=\sin x_1-\sin x_2; \end{cases}$

(6) $\begin{cases} \dfrac{dx_1}{dt}=x_1+2x_2+3\sin x_2, \\ \dfrac{dx_2}{dt}=-3x_2-x_1e^{x_1}; \end{cases}$

(7) $\begin{cases} \dfrac{dx_1}{dt}=-x_2+x_1^3, \\ \dfrac{dx_2}{dt}=x_1+\alpha x_2+x_2^3. \end{cases}$ ($\alpha \neq 0$, is a constant)

Appendix

It is difficult to solve a difference or differential equation, especially the type whose symbolic analytical solution can be get is very limited. By Matlab, the numerical solution which meets the requirements of certain accuracy can be obtained. In order to better solve the practical application using difference and differential equations, we introduce the symbolic and numerical solutions through Matlab.

A. 1 Solution of difference equations

A. 1. 1 First order linear constant coefficient difference equation

【Example A. 1. 1】 The difference equation model of population number: it is assumed that the number of the kth year population is x_k and the average annual growth rate is r, then the number of the $(k+1)$th-year is

$$x_{k+1}=(1+r)x_k, \ k=0, 1, 2, \cdots$$

It is known that $x_0=100$, and better, medium and poor natural environment correspond to $r=0.0194$, -0.0324 and -0.0382, respectively. Programmed by Matlab, observe the change of the number after 20 years.

A function is first established with respect to the variables n, r,

```
function x=sqh(n,r)
a=1+r;
x=100;
for k=1:n
  x(k+1)=a*x(k);
end
```

Call the sqh function in the command window:

```
>>k=(0:20)';% a column vector
>>y1=sqh(20,0.0194);
```

```
>>y2=sqh(20,-0.0324);
>>y3=sqh(20,-0.0382);
>>round([k,y1',y2',y3'])   % rounding
```

Observe the quantitative change trend by ploting:

```
>>plot(k,y1,'--',k,y2,':',k,y3,'*')
```

The result is

```
ans =
     0    100   100   100
     1    102   97    96
     2    104   94    93
     3    106   91    89
     4    108   88    86
     5    110   85    82
     6    112   82    79
     7    114   79    76
     8    117   77    73
     9    119   74    70
    10    121   72    68
    11    124   70    65
    12    126   67    63
    13    128   65    60
    14    131   63    58
    15    133   61    56
    16    136   59    54
    17    139   57    52
    18    141   55    50
    19    144   53    48
    20    147   52    46
```

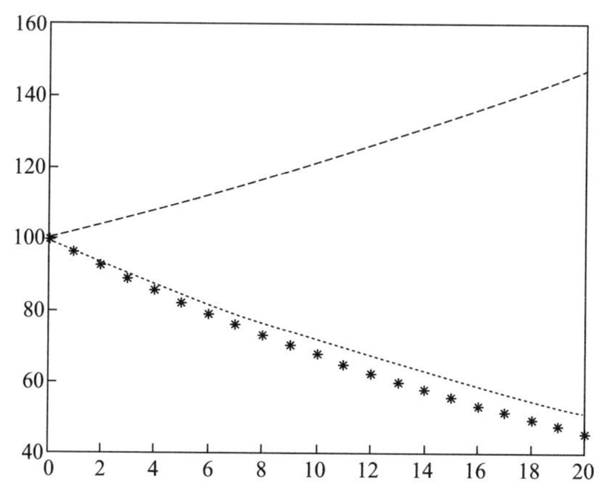

A.1.2 Higher order linear constant coefficient difference equation

【Example A.1.2】 Reproduction of annual plants: the average number of spring seeds produced by a plant is c, the proportion of seeds that survive one winter (1-year-old seeds) is b, and the proportion of seeds that survive one winter without sprouting and the second winter (2-year-old seeds) is still b. The germination rate of 1-year-old and 2-year old seed is denoted by a_1 and a_2.

Let c, a_1, a_2 be fixed, and b is a variable. The number of plants in the kth year is x_k. Obviously, x_k is related to x_{k-1}, x_{k-2}. The part determined by x_{k-1} is a_1bcx_{k-1} and x_{k-2} gives $a_2b(1-a_1)bcx_{k-2}$, that is,

$$x_k=a_1bcx_{k-1}+a_2b(1-a_1)bcx_{k-2},$$

giving a high order difference equation

$$x_k=px_{k-1}+qx_{k-2}.$$

Using Matlab with $x_0=100$, $a_1=0.5$, $a_2=0.25$, $c=10$, $b=0.18\sim0.2$,

```
function x=zwfz(x0,n,b)
c=10;a1=0.5;a2=0.25;
p=a1*b*c; q=a2*b*(1-a1)*b*c;
x(1)=x0;
x(2)=p* x0;
for  k=3:n
x(k)=p*x(k- 1)+ q*x(k-2) ;
end
```

Run the code in the comman window:

```
k=(0:20)';
y1=zwfz(100,21,0.18);
y2=zwfz(100,21,0.19);
y3=zwfz(100,21,0.20);
round([k,y1',y2',y3'])
plot(k,y1,k,y2,':',k,y3,'o'),
gtext('b=0.18'),gtext('b=0.19'),gtext('b=0.20')
```

The result is:

```
ans =
    0  100  100  100
    1  90   95   100
    2  85   95   105
    3  80   94   110
    4  76   94   115
    5  71   93   121
```

```
6   67  93  127
7   63  93  133
8   60  92  139
9   56  92  146
10  53  91  152
11  50  91  160
12  47  90  167
13  45  90  175
14  42  90  184
15  40  89  192
16  37  89  202
17  35  88  211
18  33  88  221
19  31  88  232
20  30  87  243
```

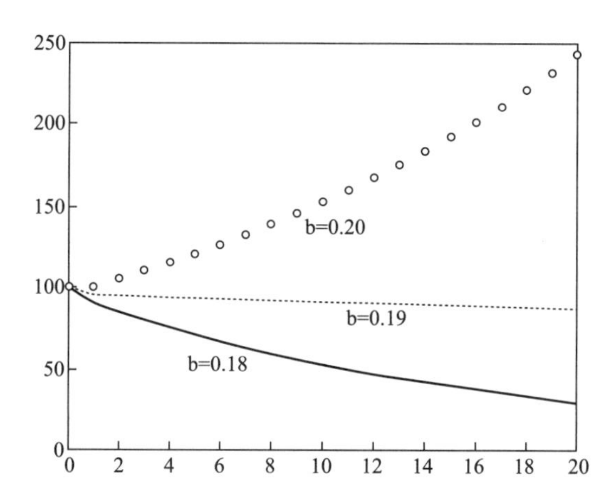

A. 1. 3 Linear constant coefficient difference equations

【Example A. 1. 3】 Solve the difference equations

$$\begin{cases} x_1(k+1)=0.6x_1(k)+0.2x_2(k)+0.1x_3(k) \\ x_2(k+1)=0.3x_1(k)+0.7x_2(k)+0.3x_3(k) \\ x_3(k+1)=0.1x_1(k)+0.1x_2(k)+0.6x_3(k) \end{cases}$$

The system of difference equations can be written as the matrix form

$$\begin{bmatrix} x_1(k+1) \\ x_2(k+1) \\ x_3(k+1) \end{bmatrix} = \begin{bmatrix} 0.6 & 0.2 & 0.1 \\ 0.3 & 0.7 & 0.3 \\ 0.1 & 0.1 & 0.6 \end{bmatrix} \begin{bmatrix} x_1(k) \\ x_2(k) \\ x_3(k) \end{bmatrix}$$

The Matlab function is

```
function x=czqc(n)
A=[0.6,0.2,0.1;0.3,0.7,0.3;0.1,0.1,0.6];
x(:,1)=[200,200,200]';
for k=1:n
  x(:,k+1)=A*x(:,k);
end
```

The following main program shows the change trend:

```
A=[0.6,0.2,0.1;0.3,0.7,0.3;0.1,0.1,0.6];
n=10;
for k=1:n
x(:,1)=[200,200,200]';
x(:,k+1)=A*x(:,k);
end
round(x)
k=0:10;
plot(k,x), grid
gtext('x1(k)'), gtext('x2(k)'), gtext('x3(k)')   %mark the curves
```

The result is

```
ans =
   200  180  176  176  178  179  179  180  180  180  180
   200  260  284  294  297  299  300  300  300  300  300
   200  160  140  130  125  122  121  121  120  120  120
```

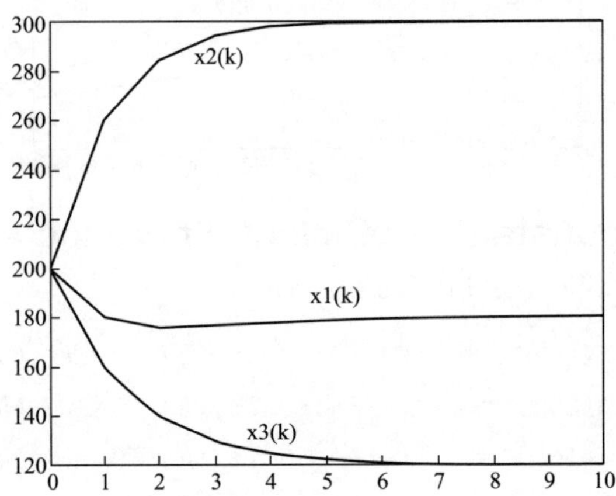

A.2 Solutions of ordinary differential equations

A.2.1 Symbolic solutions

Function: dsolve

Format: r=dsolve ('eq1, eq2, … ', 'cond1, cond2, …', 'v')

Note: v is the symbolic independent variable for a given ordinary differential equation (system) eq1, eq2, …, with the boundary and initial conditions cond1, cond2, …. Find the symbolic solution (that is, the analytical solution) r; if there is no specified variable v, the default variable is t; in the expression of the differential equation (system) eq, the letter D denotes the differential operator for the independent variable (set x): D=d/dx, D2=d2/dx2, …. The letter after the differential operator D is expressed as a dependent variable, that is, an unknown function to be solved. The initial and boundary conditions are represented by strings: y(a)=b, Dy(c)=d, D2y(e)=f and so on, giving $y(x)|_{x=a}=b$, $y'(x)|_{x=c}=d$, $y''(x)|_{x=e}=f$, respectively; if the boundary conditions are less than the order of the equation (system), then any constant C1, C2, … will appear in the returned result r; the dsolve command can accept up to 12 input parameters (including the number of equations and definite solution conditions, certainly, we can enter more than 12 equations, just putting several equations in a string). If no output variable is given, a list of solutions is displayed in the command window. When the command cannot find an analytical solution, a warning message is returned and an empty sym object is returned. At this point, the user can use the command ode23 or ode45 to solve the numerical solution of the system of equations.

【Example A. 2. 1】 Try the following command:

```
>>D1=dsolve('D2y- Dy=exp(x)')
The result is:
D1=C1-exp(x)+C2*exp(t)-t*exp(x)          %Take t as an independent variable.
>>D2=dsolve('D2y-Dy=exp(x)', 'x')
The results is:
D2=C2*exp(x)+exp(x)*(x+ C1*exp(- x))     % Take x as an independent variable.
>>D3=dsolve('(Dy)^2+ y^2=1', 's')        % Take s as an independent variable.
The result is:
D3=1
  -1
  (exp(C2*1i + s*1i)*(exp(-C2*2i-s*2i)+1))/2
  (exp(C1*1i-s*1i)*(exp(-C1*2i + s*2i)+ 1))/2
> > D4=dsolve('Dy=a* y', 'y(0)=b')                %a define condition
The result is:
D4=b* exp(a* t)                   %Take t as an independent variable.
> > sols=dsolve('D2y=-a^2* y', 'y(0)=1', 'Dy(pi/a)=0');  %two define conditions
> > D5=simplify(sols)
The result is:
D5=cos(a* t)                          %Take t as an independent variable
> > [x,y]=dsolve('Dx=y', 'Dy=-x')  %Solve a system of linear differential equations.
The result is:
x=C2*cos(t)+C1*sin(t)
y=C1*cos(t)-C2*sin(t)
> > [u,v]=dsolve('Du=u+v,Dv=u-v')
```

```
The result is
u=C1*exp(2^(1/2)*t)*(2^(1/2)+1)-C2*exp(-2^(1/2)*t)*(2^(1/2)-1)
v=C1*exp(2^(1/2)*t)+ C2*exp(- 2^(1/2)*t)
```

【Example A. 2. 2】 Give the general solution of differential equation $\frac{dy}{dx}=2xy$

Enter in the command window:

```
D=dsolve('Dy=2*x*y', 'x')
```

Available on return:

```
D=C1*exp(x^2)
```

【Example A. 2. 3】 Give the general solution of differential equation

$$\frac{dy}{dx}-\frac{2y}{x+1}=(x+1)^{\frac{5}{2}}$$

Enter in the command window:

```
D=dsolve('Dy-2*y/(x+1)=(x+1)^(5/2)', 'x')
```

Available on return:

```
D=(2*(x+1)^(7/2))/3+C1*(x+1)^2
```

that is, $y=(x+1)^2\left[\frac{2}{3}(x+1)^{\frac{3}{2}}+C\right]$

【Example A. 2. 4】 Find the special solution of differential equation $(1+x^2)y''=2xy'$ satisfying initial condition $y(0)=1$, $y'(0)=3$.

Enter in the command window:

```
D=dsolve('D2y-2*x/(x^2+ 1)*Dy=0', 'y(0)=1', 'Dy(0)=3', 'x')
```

Available on return:

```
D=x*(x^2+3)+1
```

【Example A. 2. 5】 Find the general solution of differential equation $y^{(4)}-2y'''+5y''=0$.

Enter in the command window:

```
D=dsolve('D4y- 2*D3y+5*D2y=0', 'x')
```

Available on return:

```
D=(2*C1)/25+C2+(C1*x)/5+C3*cos(2*x)* exp(x)+C4*sin(2*x)*exp(x)
```

【Example A. 2. 6】 Find the general solution of differential equation $y''+y=x\cos 2x$.

Enter in the command window:

```
sols=dsolve('D2y+y=x*cos(2*x)', 'x') ;
D=simplify(sols)
```

Available on return:

```
D=x/3-(2*x*cos(x)^2)/3+(8*cos(x)*sin(x))/9+C1*cos(x)+C2* sin(x)
```

【Example A. 2. 7】 Give the special solution of differential system $\begin{cases}\frac{dy}{dx}=3y-2z\\ \frac{dy}{dx}=2y-z\end{cases}$

satisfying initial condition $y(0)=1$, $z(0)=0$.

Enter in the command window:

```
D=dsolve('Dy=3* y-2* z', 'Dz=2* y-z', 'y(0)=1', 'z(0)=0', 'x')
```

Available on return:

```
D=z: [1×1 sym]
  y: [1×1 sym]
```

Input the next commands:

```
>>D. y
>>D. z
```

The solution is

```
y=exp(x)+2*x*exp(x)
z=2*x*exp(x)
```

A. 2. 2 Numerical solutions

1. Existing functions in Matlab

Functions: ode45, ode23, ode113, ode15s, ode23s, ode23t, ode23tb

Effect: numerical Solutions of initial value problems for ordinary differential equations.

Format: [T,Y] =solver ('odefun', tspan, y0, options, p1, p2, …)

Note: solver is one of commands ode45, ode23, ode113, ode15s, ode23s, ode23t, ode23tb.

odefun is the explicit differential equation $y'=f(t,y)$, or an equationM$(t,y)*y'=f(t,y)$ containing a mixed matrix. The command ode23 can only solve the problem of constant mixed matrix; ode23t and ode15s can solve the problem of singular matrix.

tspan is the integral interval (that is, solving interval). On the interval tspan=[t0, tf], the explicit differential equation $y'=f(t,y)$ is solved by the initial condition y0 from t0 to tf. For scalar t and column vector y, the function f=odefun (t,y) must return a column vector f of f (t,y). Each row in the solution matrix Y corresponds to a point in the returned time column vector T. To get the solution of the problem at other specified time t0, t1, t2, …, set tspan= [t0,t1, t2,…, tf] (monotonous time sequence).

y0 contains the vector of the initial condition.

options: the properties set by the parameter options (generated with the command odeset, instead of the default integral parameter). Common attributes include the relative error value Rel-Tol (the default value is 1e-3) and the absolute error vector AbsTol (the default value is 1e-6 per element).

Parameter p1, p2, p3, …, and so on are passed to the function odefun, for calculation. If there is no parameter setting, set options= [].

The basic process of solving a specific ODE:

(1) According to the laws and formulas in the subject to which the problem belongs, it is described by differential equations and initial conditions.

$$F(y, y', y'', \cdots, y^{(n)}, t)=0$$

$$y(0)=y_0, y'(0)=y_1, \cdots, y^{(n-1)}(0)=y_{n-1}$$

(2) By variable substitution $x_n=y^{(n-1)}$, $x_{n-1}=y^{(n-2)}$, $\cdots$, $x_2=y'$, $x_1=y$ in mathematics, the equations (systems) of higher order (greater than second order) are written into first order differential equations:

$$x'=\begin{bmatrix} x_1' \\ x_2' \\ \vdots \\ x_n' \end{bmatrix}=\begin{bmatrix} f_1(t,x) \\ f_2(t,x) \\ \vdots \\ f_n(t,x) \end{bmatrix}, x_0=\begin{bmatrix} x_1(0) \\ x_2(0) \\ \vdots \\ x_n(0) \end{bmatrix}=\begin{bmatrix} y(0) \\ y'(0) \\ \vdots \\ y^{(n-1)}(0) \end{bmatrix}=\begin{bmatrix} y_1 \\ y_2 \\ \vdots \\ y_{n-1} \end{bmatrix}$$

(3) According to the results of (1) and (2), the M function file odefun which can calculate the derivative is compiled.

(4) The file odefun and the initial condition are passed to one of the solvers, and the solution vector y (including y and derivatives of different orders) of ODE in a specified time interval can be obtained after running.

Remark: (1) In the function commands to find the numerical solution of the differential equation, the function odefun must take dx/dt as the output and t, x as the input.

(2) When solving the equations of n unknown functions, the unsolved equations in the M function file should be written in the form of vectors of x.

【Example A. 2. 8】 Solve the differential equation $y'=\sin x$, where $x_0=0$, $y_0=-1$. First, write the right end of the derivative expression into a function file ex2 _ 1. m:

```
function yy=ex2_1(t,x)
yy=sin(t);
```

Next, call the function directly:

```
[t,x]=ode23('ex2_1', [0,pi],-1)
plot(t,x)
```

【Example A. 2. 9】 Find the numerical solution of differential equation $y''+y=1-\frac{t^2}{2\pi}$. The initial value of the independent variable t is 0, the final value is 3π, and the initial conditions are $y(0)=0, y'(0)=0$.

The higher order differential equation is transformed into a system of first order differential equations, that is, replaced by variables:

$$x=\begin{bmatrix} x_1 \\ x_2 \end{bmatrix}=\begin{bmatrix} y \\ y' \end{bmatrix}$$

$$x'=\begin{bmatrix}x_1'\\x_2'\end{bmatrix}=\begin{bmatrix}y'\\y''\end{bmatrix}=\begin{bmatrix}x_2\\-x_1+1-\frac{t^2}{2\pi}\end{bmatrix}=\begin{bmatrix}0&1\\-1&0\end{bmatrix}\begin{bmatrix}x_1\\x_2\end{bmatrix}+\begin{bmatrix}0\\1\end{bmatrix}\left(1-\frac{t^2}{2\pi}\right)$$

In this way, the right end of the derivative expression is written as a functional program ex2 _ 2. m:

```
function xdot=ex2_2(t,x)
u=1-(t.^2)/(2*pi);
xdot=[0 1;- 1 0]*x+ [0 1]'*u;
```

Then, the existing numerical integral function is called in the main program for integration:

```
clf; t0=0; tf=3* pi; x0=[0;0]
[t,x]=ode23('ex2_2', [t0,tf],x0)
y= x(:, 1)
```

【Example A. 2. 10】 Give the numerical solution of second order differential equation

$$x^2y''+xy'+\left(x^2-\frac{1}{2}\right)y=0,y'\left(\frac{\pi}{2}\right)=-\frac{2}{\pi}$$

First of all, variable substitution:

$$z=\begin{bmatrix}z_1\\z_2\end{bmatrix}=\begin{bmatrix}y\\y'\end{bmatrix}$$

$$z'=\begin{bmatrix}z_1'\\z_2'\end{bmatrix}=\begin{bmatrix}z_2\\-\frac{z_2}{x}+\left(\frac{1}{2x^2}-1\right)z_1\end{bmatrix}$$

In this way, the right end of the derivative expression is written as a functional program ex2 _ 3. m

```
function f=ex2_3(x, z)
f=[0 1; 1/(2* x^2)-1 -1/x]*z;
```

Then, the existing numerical integral function is called in the main program for integration:

```
[x, z]=ode23('ex2_3', [pi/2, pi], [2; -2/pi])
plot(x, z(:, 1))
```

【Example A. 2. 11】 Solve the classical Van der Pol differential equations describing oscillators:

$$\frac{d^2y}{dt^2}-\mu(1-y^2)\frac{dy}{dt}+y=0,\ y(0)=1,\ y'(0)=0$$

Let $x_1=y$, $x_2=dy/dt$, then

$$\frac{dx_1}{dt}=x_2$$

$$\frac{dx_2}{dt}=\mu(1-x_1^2)x_2-x_1$$

Write the function file ex2 _ 4. m:

```
function xprime=ex2_4(t,x)
global MU
xprime=[x(2); MU*(1-x(1)^2)*x(2)-x(1)];
```

Run the following in the command window:

```
>>global MU
>>MU=7;
>>Y0=[1; 0]
>>[t,x]=ode45('ex2_4', [0, 40], Y0);
>>x1=x(:,1); x2=x(:,2);
>>plot(t,x1,t,x2)
```

We can get the figure below.

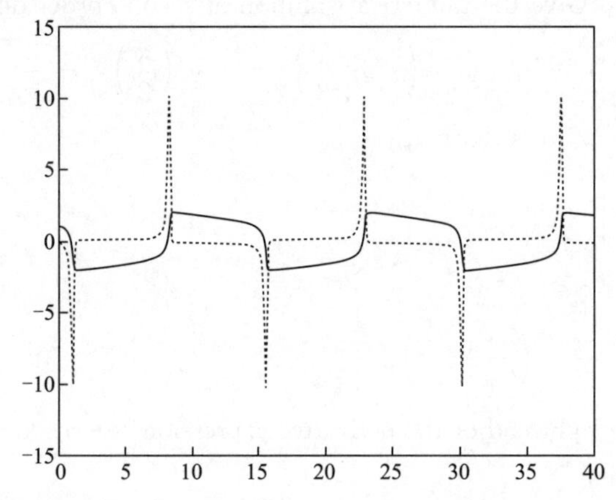

2. Other programs for numerical solutions

(1) Improved Euler program

Program name: Eulerpro. m

Format: [X,Y] =Eulerpro ('fxy', x0, y0, xend, h)

Program function: solve ordinary differential equations.

Input variable: fxy is the M function file name of the given function $y'=f(x,y)$; x0, xend are the starting and ending points; y0 is the known initial value; h is the step size.

Output variable: X is a discrete independent variable; Y is a discrete function value.

Program:

```
function [x,y]=Eulerpro(fxy,x0,y0,xend,h)
n=fix((xend-x0)/h);
y(1)=y0;
x(1)=x0;
```

```
for i=1:(n-1)
    x(i+1)=x0+i*h;
    y1=y(i)+h*feval(fxy,x(i),y(i));
    y2=y(i)+h*feval(fxy,x(i+1),y1);
    y(i+1)=(y1+y2)/2;
end
plot(x,y)
```

【Example A. 2. 12】 Solve the differential equation $y'=\sin x$, where $x_0=0$, $y_0=-1$.

Write the M function first, whose name is fxy. m.

```
function Z=fxy(x,y)
Z=sin(x);
```

The step size is 0. 1, and the call format is:

```
[x,y]= Eulerpro('fxy',0,-1,pi,0.1)
```

The result is shown in the figure:

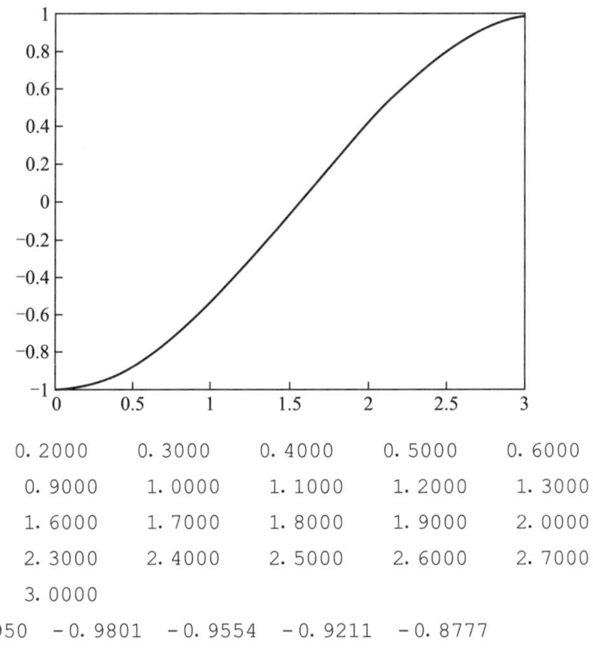


```
X =  0       0.1000     0.2000     0.3000     0.4000     0.5000     0.6000
0.7000     0.8000     0.9000     1.0000     1.1000     1.2000     1.3000
1.4000     1.5000     1.6000     1.7000     1.8000     1.9000     2.0000
2.1000     2.2000     2.3000     2.4000     2.5000     2.6000     2.7000
2.8000     2.9000     3.0000
Y = -1.0000   -0.9950   -0.9801   -0.9554   -0.9211   -0.8777
-0.8255       -0.7650   -0.6970   -0.6219   -0.5407   -0.4541
-0.3629       -0.2681   -0.1707   -0.0715    0.0283    0.1279
 0.2262        0.3222    0.4150    0.5036    0.5872    0.6649
 0.7359        0.7996    0.8553    0.9025    0.9406    0.9693    0.9883
```

(2) Runge-Kutta program

Program name: RungKt4. m

```
Format: [X,Y]= RungKt4('fxy',x0,y0,xend,M)
```

Program function: solve ordinary differential equations.

Input variable: fxy is the M function file name of the given function $y^{\wedge\prime}=f(x,y)$; x0, xend are the starting and ending points; y0 is the known initial value; M is the step number.

Output variable: X is a discrete independent variable; Y is a discrete function value.

Program:

```
function [X,Y]= RungKt4(fxy,x0,y0,xend,M)
h=(xend-x0)/M;
X=zeros(1,M+1);
Y=zeros(1,M+1);
X=x0:h:xend
Y(1)=y0;
for i=1:M
    k1=h*feval(fxy,X(i),Y(i));
    k2=h*feval(fxy,X(i)+h/2,Y(i)+k1/2);
    k3=h*feval(fxy,X(i)+h/2,Y(i)+k2/2);
    k4=h*feval(fxy,X(i)+h,Y(i)+k3);
    Y(i+1)=Y(i)+ (k1+2*k2+2*k3+k4)/6
end
plot(X,Y)
```

【Example A. 2. 13】 Solve the differential equation $y'=\sin x$, where $x_0=0$, $y_0=-1$.

Write the M function first, whose name is fxy. m.

```
function Z=fxy (x,y)
Z=sin(x);
```

The step number is 30, and the call format is:

```
[X,Y]=RungKt4('fxy',0,-1,pi,30)
```

The result is shown in the figure:

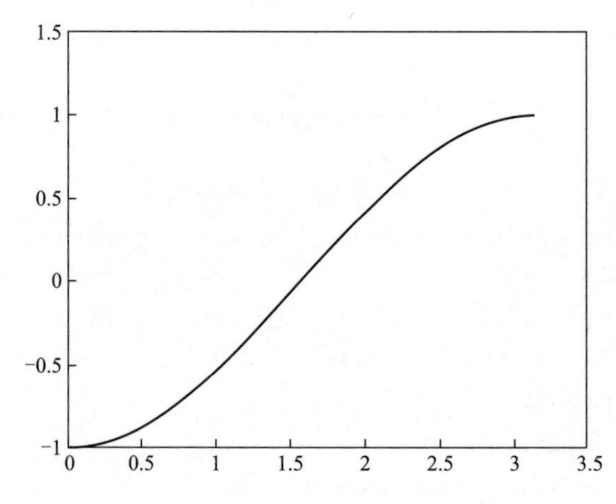

```
X= 0       0.1047    0.2094    0.3142    0.4189    0.5236    0.6283
0.7330     0.8378    0.9425    1.0472    1.1519    1.2566    1.3614
1.4661     1.5708    1.6755    1.7802    1.8850    1.9897    2.0944
2.1991     2.3038    2.4086    2.5133    2.6180    2.7227    2.8274
2.9322     3.0369    3.1416
Y= -1.0000  -0.9945  -0.9781  -0.9511  -0.9135  -0.8660  -0.8090
-0.7431     -0.6691  -0.5878  -0.5000  -0.4067  -0.3090  -0.2079
-0.1045      0.0000   0.1045   0.2079   0.3090   0.4067   0.5000
  0.5878     0.6691   0.7431   0.8090   0.8660   0.9135   0.9511
  0.9781     0.9945   1.0000
```

A.3 Exercises

1. Solve the difference equation $y_{n+1}=1.2y_n+30$ at some y_0.
2. Solve the difference equation $y_{n+1}=4y_n+y_{n-1}$ at some y_0.
3. Find the general solution of differential equation $\frac{\mathrm{d}y}{\mathrm{d}t}=ay$ and its special solution with $y(0)=b$.
4. Give the special solution of differential equation $\frac{\mathrm{d}^2y}{\mathrm{d}t^2}=-a^2y$, whose initial conditions are $y(0)=1$, $\left.\frac{\mathrm{d}y}{\mathrm{d}t}\right|_{t=\frac{\pi}{a}}=0$.

References

[1] Edwards, C. H., Penny, D. E., Elementary Differential Equations. China machine press.

[2] Elaydi, S., An Introduction to Difference Equations. Springer, 2005.

[3] 周义仓，曹慧，肖燕妮，差分方程及其应用. 北京：科学出版社，2014.

[4] Boyce, W., Diprima R., Elementary Differential Equations and Boundary Value Problems, 3rd ed. Wiley, New York, 1977.

[5] Brauer, F., Nohel, J., Ordinary Differential Equations, 2nd ed. Benjamin, New York, 1973.

[6] Braum, M., Differential Equations and Their Applications: An Introduction to Applied Mathematics, 3rd ed. Springer-Verlag, New York, 1983.

[7] Derrick, W., Grossman, S., Elementary Differential Equations with Applications, 2nd ed. Addison-Wesley, Reading, Mass., 1981.

[8] Finizo, N., Ladas, G., Ordinary Differential with Modern Applications, 2nd ed. Wadsworth, Belmont, Calif., 1982.

[9] Finney, R., Ostberg, D., Elementary Differential Equations with Linear Algebra. Addison-Wesley, Reading, Mass., 1976.

[10] Rainville, E., Bedient, P., Elementary Differential Equations, 6th ed. Macmillan, New York, 1981.

[11] Reiss, E., Callegri, A., Ahluwalia, D., Ordinary Differential Equations with Applications. Holt, Rinehart, & Winston, New York, 1976.

[12] Sanchez, D., Allen, R., Kyner, W., Differential Equations, An introduction. Addison-Wesley, Reading, Mass., 1983.

[13] Spiegel, M., Applied Differential Equations, 3rd ed. Prentice-Hall, Englewood Cliffs, N. J., 1981.

[14] Robinson, J. C., An Introduction to Ordinary Differential Equations. Cambridge University Press, 2004.

[15] Pinelas, S., Chipot, M., Dosla, Z., Differential and Difference Equations with Applications. Springer Science, Business Media, New York, 2013.

[16] Blanchard, P., Devaney, R. L., Hall, G. R., Differential Equations, 4th ed. Brooks/Cole, Cengage Learning, 2012.

[17] Tenenbaum, M., Pollard, H., Ordinary Differential Equations: An Elementary Textbook for Students of Mathematics, Engineering, and the Sciences. Dover Publications, Inc., New York, 1963.

[18] 王高雄，周之铭，朱思铭，王寿松. 常微分方程. 北京：高等教育出版社，2006.

[19] 蔡燧林. 常微分方程. 武汉：武汉大学出版社，2003.

[20] 朱思铭. 常微分方程学习辅导与习题解答. 北京：高等教育出版社，2009.

[21] 龙松，柯玲，张文钢. 大学数学 MATLAB 应用教程. 武汉：武汉大学出版社，2014.

[22] 占海明. 基于 MATLAB 的高等数学问题求解. 北京：清华大学出版社，2013.

[23] 钟益林，彭乐群，刘炳文. 常微分方程及其 Maple，MATLAB 求解. 北京：清华大学出版社，2007.